T0268995

Climate for Change

Climate for Change: Non-state Actors and the Global Politics of the Greenhouse provides a challenging explanation of the forces that have shaped the international global warming debate. Unlike existing books on the politics of climate change, this book concentrates on how non-state actors, such as scientific, environmental and industry groups, as opposed to governmental organisations, affect political outcomes in global fora on climate change. It also provides insights into the role of the media in influencing the agenda.

Critical of the predominance of state-based regime theory in the explanation of international environmental cooperation, the author makes a strong case for the centrality of non-state scientific, environmental and industry groups, as well as the mass media, to explanations of how the climate regime was formed and has evolved. The book draws on a range of analytical approaches to assess and explain the influence these non-state actors have brought to bear on the course of global climate politics. It explores the benefits of a theoretical perspective that fuses insights from international political economy with those of transnational relations in order to capture more adequately the different dimensions of the power of non-state actors in global environmental politics.

The book will be of interest to all researchers and policy-makers associated with climate change, and will be used in university courses in international relations, politics and environmental studies.

Climate for Change

Non-state Actors and the Global Politics of the Greenhouse

PETER NEWELL

University of Warwick

CAMBRIDGE
UNIVERSITY PRESS

CAMBRIDGE UNIVERSITY PRESS
Cambridge, New York, Melbourne, Madrid, Cape Town, Singapore, São Paulo

Cambridge University Press
The Edinburgh Building, Cambridge CB2 2RU, UK

Published in the United States of America by Cambridge University Press, New York

www.cambridge.org
Information on this title: www.cambridge.org/9780521632508

First published 2000
This digitally printed first paperback version (with corrections) 2006

A catalogue record for this publication is available from the British Library

Library of Congress Cataloguing in Publication data
Newell, Peter (Peter John)
    Climate for change: non-state actors and the global politics of the greenhouse / Peter Newell.
        p.   cm.
    Includes bibliographical references.
    ISBN 0-521-63250-1 (hb)
    1. Greenhouse effect, Atmospheric.   I. Title.
    QC912.3.N49   2000
    363.738′7457 – dc21   99-087673

ISBN-13 978-0-521-63250-8 hardback
ISBN-10 0-521-63250-1 hardback

ISBN-13 978-0-521-02123-4 paperback
ISBN-10 0-521-02123-5 paperback

Contents

Acknowledgements

As is often the case, this book started life as a PhD thesis. It was completed at the University of Keele between 1993 and 1996. Firstly, therefore, I would like to acknowledge the invaluable assistance of Matthew Paterson and David Scrivener, the two supervisors who guided me through the logistical and psychological battlefield that is doing a PhD.

Thanks are due to the numerous colleagues who commented on earlier drafts of the chapters that make up this book. John Vogler and Brian Doherty, the examiners of my thesis, gave me an immense amount of positive and constructive feedback. In addition, I am grateful to the anonymous reviewers of the proposal for this book who made a number of useful suggestions, and to Matt Lloyd at CUP, for guidance and encouragement. Finally, I would like to extend my appreciation to all those NGOs and government officials who spared the time to complete questionnaires, be interviewed or allowed me to use their archives.

The financial support of the following is also gratefully acknowledged: the British International Studies Association, the Gilbert Murray Trust Fund, the Polehampton Trust and the Department of International Relations at the University of Keele, the Department of Politics and International Studies at Warwick and the MacArthur Foundation.

The research for and writing up of this book have taken me on a journey from Keele to Warwick to Brighton. At each stage I have been lucky enough to form great friendships that I would like to acknowledge briefly. At Keele, Ben Seel, Matt Paterson, 'Sparky' Bedwell, Fiona Candelin, Glyn Williams, Rosarie McCarthy, Johnny Mac(Millan), Paul (Alty) Johnson and too many others to list individually, all deserve thanks. Philippa Bell, in particular, was a wonderful companion. At Warwick, the group of individuals collectively known as the 'geezers' (Richard Devetak, Charlie Dannreuther, Rohit Lekhi, Jane Booth and Ben Rosamond) have been a source of entertainment and life beyond work as well as providing greatly valued friendships. At IDS, the band, the football team and the rest of the staff have all been great people to be around, and the secretarial support of Linda Bateman has been critical in keeping to deadlines.

Rahul Moodgal deserves special mention for being a wonderfully caring human being. His friendship means everything to me. The love and support of Bridget Allan have been an immense source of strength in this endeavour. Huge thanks are due to all these people. The helter-skelter fortunes of the mighty

Seagulls (Brighton and Hove Albion FC) have made the ups and downs of writing a book look like a smooth ride. May future times be less turbulent and a little more successful.

I dedicate this book to my parents and my sister for their loving support for everything I have done.

Peter Newell
Brighton, 2000

1

Politics in a warming world: introduction

1.1 Introduction

To understand the ebb and flow of the climate change issue in national and international contexts requires an appreciation of the way in which political power is exercised by different groups in pursuit of their aims and objectives. (O'Riordan and Jordan 1996:78)

This book is concerned with explanations of the content and formulation of international climate policy; the way in which one might account for the efforts of the international community to engage with the question of human interference with the global climate system. In an attempt to understand the nature and scope of international climate policy from a new perspective, discussion focuses on the political impact of four sets of non-governmental actors[1] whose importance has not been conceptualised in a developed manner in the literature on global warming. The terms non-governmental and non-state actors are used interchangeably throughout the book and refer to actors that are not officially part of national government.

The four groups of non-state actor looked at in relation to the politics of global warming are Working Group 1 of the IPCC (Intergovernmental Panel on Climate Change), the mass media, the fossil fuel lobbies and environmental pressure groups. These non-governmental actors in particular, have been chosen as a means by which to challenge predominant explanations in the literature on global environmental politics, which generally lack analysis of these actors. Hence the purpose of this book is twofold. Firstly, it seeks to redress the imbalance in the international relations literature on global environmental politics towards state-centric analysis of 'regimes'[2] as the key location for explanations of political outcomes. It does this by focusing upon the importance of sub- and trans-state non-governmental actors. In so doing, the need for inter-state analysis is not negated, rather an argument is made that analysis of NGOs in the politics of global warming raises important challenges to conventional thinking about the sources of political outcomes in global environmental politics.

---

[1] For a lengthy discussion of the definitional issues that attend any attempt accurately to describe NGOs, see Willetts (1993).

[2] The definition of an international regime applied throughout the book is the widely used Krasner (1983:2) definition of 'sets of implicit or explicit principles, norms, rules and decision-making procedures around which actors' expectations converge in a given area of International Relations'.

In part, it remains wedded to the regime project of explaining regime formation, maintenance and change. But it focuses far less upon the institutions themselves and the way in which they can influence state behaviour, and looks instead at what have, until now, been considered contextual factors or externalities.[3] Hence it does not seek primarily to challenge the authority of the claims made about inter-state bargaining (although appropriate criticisms are levelled), but rather the ability of regime approaches to account for the range of influences upon international climate change policy, without attempting to include analysis of domestic and transnational non-governmental actors and the influence they may bring to bear upon the course of international politics.[4]

The second key purpose of the book is to embellish existing explanations of the political dynamics at work in relation exclusively to climate change. The argument is developed that the politics of global warming require broader approaches to understanding international cooperation than are provided by regime theoretical accounts. Analysis of the interaction of actors both inside and outside narrowly defined institutional settings can contribute towards such an understanding. The book therefore attempts to show how analysis of NGO actors may be particularly pertinent to explaining the politics of global warming.

Looking at the issue of global warming in specific terms enables a better understanding of 'how issue areas shape the relative power of NGOs' (Haufler 1995:110). This is important in respect of the first goal of the book (examining how NGOs are influential). Case study approaches are also necessary in order to explore fully the power relations that characterise a specific issue area such as global warming (the second goal of the book). Problem structures differ according to the issue in question and generalising accounts of international cooperation need to be made more sensitive to this. As Snidal notes, 'analysis of the formation and development of international political regimes cannot be studied without an appropriate understanding of the strategic structure of the underlying issue area' (Snidal 1985a:941).

1.2   Why these actors?

Non-governmental actors per se have not received extensive attention within the discipline of international relations (IR) (Willetts 1993). Besides general references to their importance, there has been 'little emphasis on theorising NGOs as non-state actors in the IR literature' (Elliott 1992:1), nor empirical documentation

---

[3] In many cases the assumptions are implicit (Smith 1993) given that, as Young (1989a:9) notes, 'much of the growing literature on international regimes consists of descriptive accounts of specific institutional arrangements'.

[4] It would appear to be fair to do this on the basis that regime theory sets its goal as explaining behaviour in a 'given issue area of International Relations' (Krasner 1983:2).

of their activities. There are a number of explanations for the lack of academic attention to the impact of NGOs at the global level. One is that NGOs are not considered to be powerful in the way that states are. They do not generally possess many of the resources that are traditionally considered to confer power upon actors on the global stage. Such resources are assumed to be an ability to mobilise violence, to control territory or a population, and economic power (Goldmann and Sjöstedt 1979; Willetts 1993). Under such a narrow definition of power, the place of states is privileged over other actors (Elliott 1992). The wider point is that NGOs are not thought directly to address the sorts of security issue that are traditionally of concern to international relations specialists. Moreover, the arena of foreign policy-making is considered to be largely immune to non-state pressures, as one of the least open sectors of government policy. That states set the boundaries within which these actors operate, and that most NGOs are too 'weak' to have an impact on world affairs are further perceptions that explain the neglect of NGOs in IR (Haufler 1995:96). Willetts (1982:18) uses a quote from Reynolds and McKinlay to make this point: 'As far as INGOs are concerned it is evident that the consequences of the activity of many of them are trivial. ... They may serve in some degree to alter the domestic environment of decision-makers, but with some exceptions their effect either on capabilities or on objectives is likely to be minimal, and in no way can they be seen themselves as significant actors.' NGOs are thought to matter only in issues of 'low politics', and even then only on terms and conditions established by states (Waltz 1979:94–5). Further, the scope, scale and variety that characterise the NGO phenomenon provide any potential researcher of their importance in international politics with a daunting task.

Given this background, there are few precedents for the study of NGOs in global politics. Despite the attempt by transnationalist/complex interdependence scholars (Keohane and Nye 1977; Nye and Keohane 1972) to put non-state actors onto the intellectual map, analysis of NGOs is not yet an accepted feature of the international relations discipline. It is perhaps especially ironic that IR thinkers from this transnationalist school who sought to place the importance of NGOs on the agenda of the discipline, lost sight of their importance when they came to look at regimes (Putnam 1988; Vogler 1995). Hence whilst Nye and Keohane (1972:x) decry the fact that transnational actors have 'often been ignored', when it comes to regime analysis their own work lacks any attempt to integrate NGOs (Keohane 1995). Risse-Kappen (1995:7) notes in this respect that the first debate on transnational relations in IR 'essentially resulted in confirming the state-centred view of world politics'.

Hence whilst there is, in some quarters, acknowledgement of the role of NGOs as political actors (Caldwell 1990; Carroll 1988), there have been few attempts to 'ascribe to them any major importance in determining international political outcomes' nor, more importantly, to 'acknowledge a need to rethink models of International Relations' (Elliott 1992:10). Regime analysis is largely silent on the role of non-state groups at the global level (Risse-Kappen 1995). Young's defining text, *International Cooperation* (1989c), devotes less than one of 236 pages to their importance.[5] Even here the discussion centres on the way in which 'international regimes ... give rise to non-governmental interest groups' (Young 1989:78), and not the way in which NGOs may shape the institutions and practices of the regime.

Elsewhere NGOs are emphasised in order to draw attention to ways in which they may strengthen states' capacity to cooperate (P. Haas 1990a; Young 1989c). It is argued that NGOs strengthen and reinforce regime functions by performing 'watchdog' (ibid.; Wettestad 1995) functions in helping to ensure compliance, and from applying pressure on 'laggard' states (Porter and Brown 1991). As actors in their own right, however, NGOs remain unimportant in these conceptualisations. Attention to the ways in which NGOs may bring about changes in the behaviour of states, or set agendas, is lacking. In Young's conceptualisation, NGOs are reactive to agendas already established by regimes and the governments party to them. It is assumed, moreover, that whilst 'powerful groups do sometimes succeed in exercising considerable influence over the shape of social institutions at the domestic level', at the international level the key actors are always 'dominant states or coalitions' (Young 1989c:69). The assumption is that states dictate the terms of participation and influence for NGOs (Raustiala 1996). There is very little sense in which the relationship might run both ways.[6]

More recent work on NGOs falls into the same trap. Arts (1998:56) argues that

there are of course mutual connections and interactions, but it would go too far to see the NGO–state relationships at global level as one characterised by interdependence. States are definitely dominant in the international arena and, moreover, their governments are the formal policy and decision-makers. Therefore their dependence on NGOs is generally quite limited. Whereas states at national level have recently handed over formal competencies to private players in accordance with the neo-liberal ideology, such is hardly the case at global level.

Such an approach reduces the complexity of NGO power to a narrow range of impacts on formal policy outcomes produced by those within the policy arena. Less determinable patterns of influence are rejected in favour of 'hard evidence' of outcomes that will always be easier to equate with state intervention and power. For Arts (1998), power is exercised by and upon those within the policy arena only. Indeed he claims that all influence is conditional on a 'friendly' government carrying an NGO proposal on its behalf. 'This is a prerequisite for any NGO influence' (ibid.:231). Anything other than global level activity (narrowly-defined) is excluded from Arts's analysis.

---

[5] For Young (1989c:53) NGOs 'seek to ameliorate well-defined [presumably not by them] problems rather than assume any major role in restructuring the institutional arrangements prevailing in international society'. Their significance therefore derives from the contribution they make 'toward the development of a richer texture of institutional arrangements' (ibid.:54).

[6] One possible exception to this is the work of Peter Haas. Much more is said about this in Chapter 3 of this book.

This project inverts, then, the conventional understanding of state–NGO relations, where the latter are defined and constrained in their influence by the state in a linear and unproblematic way by taking NGO actors as the starting point. It explores the relations between NGOs and the state in a way that is sensitive to the power of both. It goes beyond seeking to determine which is more influential: the state or NGOs, as if they are not interdependent. By focusing on NGOs, the role of the state in restraining or enabling the power of these actors is not downplayed. Rather, it is a dynamic process rather than a static one-way flow of influence from non-state actors to the state that is the subject of this enquiry.

Strictly non-state-actor analysis, in abstraction, would perhaps not develop our thinking very far. The forms of NGO influence looked at here are in many ways defined by the state and the impact of NGOs upon *international* politics; the forms of leverage of different groups of actor in relation to the state. This is the most appropriate way of emphasising their importance to traditional scholars of international relations, who have become accustomed to overlooking non-state actors and privileging the state in their analysis. As a first step towards a more meaningful inclusion of the importance of non-governmental actors in explanations of international environmental politics, this strategy is justified.

Each chapter is intended to assess ways in which the political role of these groups may be important for explaining the international politics of climate change. This prompts discussion of wider questions about 'influence' as a political concept and the networks of influence of which non-state actors are a part. Unlike the argument of some writers that the policy impact of transnational actors does not vary systematically with the types of actor involved (Risse-Kappen 1995), it is argued here that the political influence of different actors needs to be thought about in distinct ways. In relation to each group of actors considered, the relevant chapter reviews the influence of a broad range of actors in that sector of non-state activity. The breadth of analysis of these various actors, combined with the brief to explore the politics of global warming, limits the application of the study as an insight into the functions of these actors in broad terms. But it does say something useful about their importance to the policy debate on global warming. Coverage of a range of non-governmental actors is considered desirable in order to demonstrate the different types of political influence that are at work in global climate politics. Further, within each group of actors explored in the book, a diversity of players are touched upon to show how seemingly similar groups can have very different forms of influence.

The particular groups of non-governmental actor have been chosen for different reasons in each case. It is sufficient here merely to review the principal reasons for their inclusion in the book.

In the case of the *scientific community*, the focus of Chapter 3 is Working Group 1 of the Intergovernmental Panel on Climate Change (IPCC). The work of Peter Haas on 'epistemic communities', which focuses on the role of knowledge-based scientific communities in enhancing international cooperation, is employed extensively in this chapter. It is one of the few attempts by regime scholars to consider in any detail the impact of NGOs upon international policy. Global warming, in particular, is characterised by a dependence upon scientists to define the responses to the issue (Skolnikoff 1990), such that the scientific community potentially has a key role to play in the problem's resolution.

With regard to the *mass media*, very little has been written on the nature or political impact of the mass media's coverage of environmental issues, and even less their coverage of global warming. There is nothing in the literature on global environmental politics on what analysis of the mass media might bring to explanations of policy. Chapter 4 seeks to redress this deficit by drawing on work in media studies to show how the media can influence the course of political events by framing debate in a particular way. In regime terms, this can be thought of as looking at the 'stories which generate problem-setting and set the directions of problem-solving' (Jönsson 1995:211).

In relation to the discussion both of the influence of the scientific community and of the mass media, emphasis is placed on the way in which 'control' of knowledge and meaning is an important power resource. It brings to the fore discussion of the importance of the perception and interpretation of problems, and the actors that are in a position to inform these. It attempts to go beyond an assessment merely of how 'institutions establish the range of discourse and available options' (ibid.:715) and looks instead at how non-governmental actors have a role to play in framing policy debates.

The role of industrial groups in general has received scant attention in the international relations literature, and consideration of the political role of the *fossil fuel lobbies* is equally lacking in the literature on global warming. By drawing out connections between the interests of these lobbies and the interests of states in relation to the climate issue, Chapter 5 posits three levels of influence in relation to the power of the lobbies, two of which relate not to outward and observable lobbying, but to the power of their presence in other areas of government policy, and to the structural influence that they are in a position to exert over states' climate policy strategies. The study of the influence of key corporate actors informs our understanding of the degree of manoeuvrability open to states in their deliberations on climate policy. The neglected issue of regime prevention features highly in this chapter.

The final 'actor' chapter (Chapter 6) deals with *environmental pressure groups*, which have received far more attention in the literature on global environmental politics. The chapter is centrally concerned with those pressure groups that have devoted considerable lobbying energy to the issue of climate change at the international level. It explores the opportunities and constraints that environmental NGOs have been able to exploit, or have been forced to adapt to, in their efforts to mobilise action on the issue of climate change. The potential for influence is shown to differ widely according to the nature of the group in question and the context in which it is operating.

Each of the chapters includes an analytical breakdown of the policy process, with the exception of Chapter 4 on the mass media, where only a focus on agenda-setting is appropriate.[7] This is intended to ensure that the analysis is sensitive to the multifaceted and dynamic nature of political influence: the way it changes over time in different situations. The breakdown used is (1) agenda-setting, (2) negotiation-bargaining and (3) implementation, and is broadly compatible with similar formulations by Boehmer-Christiansen (1989), Haas, Keohane and Levy (1993), Young (1989a) and Young and Osherenko (1993).[8] There is some overlap between the different stages identified. All three stages can exist simultaneously so that, for example, whilst negotiations are proceeding on the eventual form of a protocol at the international level, convention obligations are still being implemented nationally, and interest groups are pressing upon government departments their preferred proposals for any protocol that may emerge internationally. Broadly speaking, however, *agenda-setting* refers to the phase of problem and interest definition in response to an issue, principally at the national level in the first instance. It describes the process where interested parties are called upon, or mobilise themselves, to participate in the debate on how a government should respond to a 'new' problem. Temporally, this stage covers the whole preglobal negotiation period.

Negotiation-bargaining refers to the stage of the policy process when agreement has been reached on the need for internationally coordinated response mechanisms. This phase is characterised by bargaining over suitable settlements and how burdens should be distributed between states. This is the stuff of regime theorising. Finally, there is the *implementation* stage, which is often neglected in writing on international cooperation (Greene 1996), when policies are put in place to meet obligations agreed upon in international fora and treaties are ratified. The focus once again is primarily on the national level.

Each chapter contains a short section on *structural factors* and *bargaining assets* particular to the group of actors in question. This serves to focus attention on the particular situation of this group of actors in the debate, and deals with the positional influence of the group at a general level. It offers a framework for understanding the specific forms of influence, which are then drawn out in the main body of the chapter.

The two terms might be differentiated in the following way. *Structural factors* are, for example, the relations of dependency that exist between the state and the suppliers of energy – the fossil fuel utilities – or states' dependency on the knowledge generated by scientific experts. Structural power in this sense relates to Susan Strange's use of the term: the power to establish the context within which others make decisions (Strange 1988). It also describes enduring positional influence, as opposed to temporary or fortuitous influence. *Bargaining assets* refer to points of leverage that groups are able to use to advance their position with governments. Examples include environmentalists' claims that they represent public concern about the environment,

---

[7] Unlike the other actors analysed in this book, the mass media are not prominent players during the negotiating or implementation stage, so the regime breakdown is less useful in this instance. The analysis in Chapter 4 focuses on the broadly conceived agenda-setting stage of the policy process.

[8] Though agenda-setting for these writers takes place at the international level.

or the media's access to public audiences, which confers significance on the way in which they represent the global warming issue. The distinction is not absolute, but it serves to clarify the structural relationships and points of leverage that provide a context for understanding specific forms of influence that operate in the politics of global warming.

1.3   Why global warming?

The issue of global warming[9] has been chosen for a number of reasons. Firstly, the regime is still at what may be considered an embryonic stage of development. It therefore provides an opportunity to offer a more refined account of the politics that will enhance, or militate against, future efforts to grapple with the problem.

Secondly, global warming and the political and economic problematics that underlie it simultaneously provide one of the most interesting, but also complex, environmental problems facing the international community. Interesting, because of the political challenges that are thrown up in terms of the scale of international cooperation that will be required to address the threat. Complex, because of the way in which global warming is part of, and interacts with, so many other issues on the international agenda, such as deforestation, international aid and a series of North–South relationships.

Global warming is unique in a number of senses compared with other environmental problems the international community has faced. The *problem structure* of global warming gives rise to particular sets of political relations that need to be understood in a focused and issue-specific way. Problem-structural approaches emphasise how the characteristics of an issue help to determine the probability of regime formation and change (Breitmeier and Dieter Wolf 1995; List and Rittberger 1992). Although this notion is not a new one and borrows from Lowi's (1964) work, the approach has not been emphasised in the literature on global environmental politics, though O'Riordan and Jaeger (1996) briefly discuss the idea in their work on climate change.

Skolnikoff (1990) implicitly subscribes to a problem-structural approach by identifying four special features of the global warming issue that together make it a particularly intractable issue to resolve. Firstly, there is the fact that the problem is inextricably related to so many other issues on the global agenda. Secondly, the difficulty of estimating the physical and socioeconomic impacts of the problem discourages a sense of urgency in dealing with the problem. Thirdly, the truly

---

[9] The terms 'global warming', 'climate change' and 'global climate change' are used interchangeably in this book to refer to the same scientific phenomenon. It is acknowledged, however, that the terms are politically loaded. Environmental pressure groups and the mass media, for example, seem to prefer the term 'global warming', because the term has a more emotive or dramatic resonance. The scientific community and the fossil fuel lobbies seem to prefer the term 'global [climate] change', because it sounds less alarming.

global nature of the problem requires the cooperation of a diverse range of political actors and interests, complicating the likelihood of finding solutions acceptable to all. Finally, political responses to the issue of global warming are argued to be dependent to a greater degree than other issues upon the advice of scientists. For Skolnikoff (ibid.) this may have the effect of delaying the prospect of meaningful political action, since consensus within scientific communities is reached only very slowly.

More than most other environmental issues, global warming goes to the heart of the modern industrial economy. Energy, especially reserves of cheap fossil fuel energy such as coal, oil and gas, drives economic growth in the contemporary global economy. Most problematically, the largest and most powerful states and regions in the global economy (the US, Europe, Australia and China) are sustained by the profligate use of cheap and readily available reserves of these resources. Hence unlike the issue of ozone depletion, with which it is often compared (Benedick 1991c; Rowlands 1995; Sebenius 1991), global warming relates to basic patterns of production and consumption, and potentially their transformation. As Rowlands (1995) notes when comparing the two issues, confronting global warming is about dissipating business and not different business, less about the replacement of offending substances or the creation of substitutes (as is the case with ozone depletion) and more about reduced output and changes in entrenched patterns of behaviour. The scale of resistance and inertia that an effective, long-term solution to global warming needs to confront are vast and unlike anything witnessed to date in addressing other environmental problems. Because of this, Lunde (1995:52) notes that 'global warming has a stronger scent of "high politics" than any other environmental problem'.

This leads to the third answer to the question 'why global warming?', and to the question 'why non-governmental actors and global warming?' Given the nature of the interests that are aligned against further action on global warming, the threat that global warming poses to the conventional operation of industrial economies and governments' reluctance to face up to these challenges, analyses of non-governmental actors becomes pertinent as a means of locating the potential sources of change and catalysts to government action. The scientific community, environmental groups and the mass media, by raising public awareness and putting pressure on politicians, can create momentum which, in relation to other environmental issues, has been successful in bringing about policy changes at government level. The activities and pressure for change that the actors examined in this book are capable of generating, may play a critical role in determining the nature and degree of policy response that is developed at the international level.

Many of the key obstacles that analysts have identified as standing in the way of further resolution of the global warming issue can also be better understood from an NGO perspective. Hahn and Richards (1989:446) note, for example, that 'A coordinated strategy aimed at prevention would require both a much greater consensus on the scientific aspects of the problem or a much greater level of public concern than currently exists.' Assuming they are correct, actors that may be in a position to activate public demands, or contribute to policy-relevant scientific understanding become central to an understanding of the surmountability of these obstacles.

1.4 Methodology

Part of the problem in attempting to construct an analysis that captures the political dynamics at work in environmental politics is that, as Elliott (1992:6) notes, 'Assessments of influence and success often rely on the perceptions of NGOs which may overstate the case, or of governments which may wish to play down NGO influence or to claim successful NGO initiatives for themselves.' One way of tackling this difficult problem is to make clear the limits of simply recording the opinions of key actors through questionnaires and interviews, and to explore the issues through conceptual and theoretical lenses in order to obtain a broader picture. It is not enough to rely on the opinions of the actors involved, or observed accounts of actor interaction.

The analysis in this book goes beyond an examination of politics within the formal decision-making circle. Direct observation benefits from integration with analysis of a more conceptual nature that addresses issues of non-participation, exclusion and agenda-setting. Such a conceptual framework is provided by the work of Bachrach and Baratz (1962) and Crenson (1971); what is often referred to as the second-dimensional (non-observable) approach to power (Lukes 1974). Accounting for the political origins of inaction, which is part of the brief here, is an imprecise exercise prone to a range of charges from positivist policy analysts.

Criticism of an approach that explores 'non-issues' and tacit power is grounded in negation of the idea that there necessarily is an explanation for inaction. For pluralist analysts of the policy process (Dahl 1961, 1963) there are non-issues only where there is non-interest. In other words it is assumed that a particular policy course is not pursued or a particular type of issue not raised because there is insufficient organised political interest on that issue. This is the view that 'sources of political neglect are not themselves political' (Crenson 1971:130).

For Dahl (1963:52), the major methodological problem with second-dimensional approaches is that 'seemingly well-placed observers can be misled by false reputations; they may attribute great power where little or none exists'. Yet if influence is perceived by decision-makers and can be said to have informed their decisions, then regardless of whether that influence can be directly and unquestionably attributed to an actor, it nevertheless helps to account for an outcome, which is the goal here. The section entitled 'Reconfiguring political influence' in Chapter 2 deals with these issues at length. This section only explores the methodological issues involved in researching influence in a way that goes beyond the mere association of stimulus and response.

The approach here is not to reject analysis that focuses strictly on actions, but to draw attention to the importance of tacit, less observable influence. The reputation for being powerful may of itself obstruct action on an issue, but that is not to say that reputation may not also have been reinforced by acts or demonstrations of power. It is not a question of trying to demonstrate the superior explanatory value of the observable over the non-observable. Rather the issue is the extent to which perceived power, not *necessarily* based on actions, and not *necessarily* supported by *acts* of power, influences decision-making. It may be the case, for instance, that the exercise of indirect influence permits actors to refrain from exerting direct influence (Crenson 1971). These approaches may be particularly useful in the light of Knudsen's (1979) argument that the secrecy and inaccessibility that characterises the majority of political interactions requires a broad methodological approach. He notes, 'To get at influence structures in international politics the researcher thus has little choice but to employ indirect methods' (ibid.:98).

Given this, the issue is how to research perception. As Allison (1971:181) notes, 'Documents do not capture this kind of information. What documents do preserve tends to obscure, as much as enlighten.' Diaries offer one insight into the perceptions of leaders and government figures. However, given the contemporary nature of the subject matter here, few diaries, biographies or autobiographies have been written (though Mintzer and Leonard's (1994) edited volume offers the more formal insights of leading negotiators and officials). In any case, such sources would be unlikely to make perceptions of the power of different NGO actors a major part of their discussion, even if the occasional passing and unwitting piece of evidence provides useful insights into the perceptions of the key actors involved. For Allison (1971), perceptions are gleaned from access to participants in decision-making. This implies the use of questionnaires, interviews and discussions with participants in the policy process.

Interviews with NGOs and decision-makers involved in the policy process, and attendance at key meetings of prominent actors have been used to build up a picture of NGO influence. Attendance at these meetings permitted me to undertake recorded interviews with assorted non-governmental and state actors. The use of interviews and questionnaires was intended to help fuse strictly conceptual analysis of political influence, which may be prone to abstraction, with the minutiae of the global warming story, as it is experienced by the actors involved. Face-to-face and telephone interviews were conducted with participants in the international climate change negotiations. Telephone interviews provided the opportunity to speak to people who could not predict, even on a day-to-day basis, when they would be free to be interviewed. A number of postal questionnaires were sent out to obtain feedback from key players who were either difficult to get hold of, or were too busy to be interviewed. They provided additional insights that enrich and broaden the analysis presented in this book. Of particular significance in this respect is the fact that the questionnaires enabled input from many actors whose perspectives are often left out of academic analysis (such as environmental groups from the southern hemisphere, which are often underrepresented at international meetings). The questionnaires provided a useful and comparable way of assessing different groups' perceptions of their political influence, given that the questionnaires were structured along similar lines. They offered a comparative insight into how the respondents viewed the influence of other groups. Given the relative nature of the concept of influence – its presence is always contingent on the activities and perceived capacities of other groups in the process – this cross-referencing to the situation of other groups is particularly important.

The use of questionnaires and interviews helps to get beyond dependence on institutional observation, where key observable 'events' within formal fora are used as explanations of outcome, by opening up channels through which to explore actors' perceptions. They help to flesh out, for example, accounts of why things are not done, why particularly policies are not adopted, and therefore offer a useful supplement to records of direct and formal bargaining.

In relation to my work on the role of environmental pressure groups, some time was also spent working at organisations actively campaigning on the climate change issue: Friends of the Earth UK and Climate Network Europe in Brussels. This form of participant observation[10] provided its own insights into the limitations and opportunities that exist for NGOs seeking to influence global climate politics. It facilitated access to leading campaigners and to campaign meetings. I was also able to obtain press releases and gain access to media archives and a vast range of other useful materials. The time spent working at a key branch of the global CAN (Climate Action Network) network, the principal umbrella organisation for environmental non-governmental input into the climate policy debate, enabled a degree of access to leading campaigners and private documents, and allowed me to partake in strategy discussions that have supplemented in important ways, the findings of this book by providing an insight into the day-to-day lobbying and other activities of one of the organisations that are the subject of the research. In circumstances such as these there develops a 'contractual relationship' between groups and the researcher (Melucci 1989:236). The danger is that the researcher becomes too intimately involved with the group and identifies with it. Yet this is only a problem if one takes the view that a social group can be objectively or 'scientifically' assessed and that distance from the subject breeds better research. The counter view is that it is impossible to do research 'uncontaminated by personal and political sympathies' (Becker 1970a:15). It is also the case that being closer to the activities being studied provides valuable insights that are not otherwise available to the academic who observes from the sideline.

1.5   Scope

This book principally covers the period from the emergence of the issue upon the international political agenda in 1988 through to the negotiations leading to the agreement of a protocol in Kyoto in December 1997. Although reference is made to the discussions at the Fourth Conference of the Parties in Buenos Aires in

---

[10] For more on the methodological difficulties involved in participant observation see Becker (1970a).

November 1998, the bulk of examples are drawn from the period 1988–97 and the political events that characterised the negotiations during that period.

1.6   Background: emerging themes in global climate politics

In plain language the bottom line of the convention is not climate change but what to do about it, when and at what cost. (Michael Zammit Cutajar before the World Climate Programme meeting 1993, quoted in Nilsson and Pitt 1994:41)

This section reviews some of the more significant and defining issues and events in the political history of the issue of global climate change to date, in order to provide some background to the analysis that follows.

Negotiating blocs

Diversity of interests lie at the heart of climate politics. There are a number of notable coalitions of countries within the climate negotiations representing a range of perspectives on the issue.[11] At one end of the spectrum there is the Organisation of Petroleum Exporting Countries (OPEC), which includes states such as Saudi Arabia and Kuwait that are dependent on exports of fossil fuels. This group of states has resisted any controls on the emission of greenhouse gases by continually drawing attention to the economic costs that will be incurred by action to reduce greenhouse gas emissions, and hence they may be said to constitute a 'veto' coalition in the negotiations (Porter and Brown 1991). Their strategy has been to hold up negotiations as much as possible by referring to rules of procedure, disputing the minutiae of draft texts and fiercely resisting the input of environmental NGOs during the negotiations. These states have been in close contact with the fossil fuel lobby groups (the subject of Chapter 5 of this book), which have assisted them by supplying strategic information and political support in the negotiations.

At the other end of the spectrum there is the Alliance of Small Island States (AOSIS), the states that are most threatened by global climate change and calling for the most drastic climate change abatement action. These states drafted a protocol calling for a 20 percent reduction in CO_2 emissions by the year 2005 (otherwise known as the Toronto target). The AOSIS states have proved to be compatible political partners with environmental groups such as Foundation for International Environmental Law and Development (FIELD) and Greenpeace in attempting to catalyse international action on the climate issue. These groups have been able to supply scientific and technical legal expertise in support of the AOSIS states' negotiating position.

---

[11] For more on the early negotiating blocs in the climate talks see Paterson and Grubb (1992). For more recent developments see Newell (1997, 1998) and Newell and Paterson (1996).

Europe

Between these two extremes lie the majority of the other key players in the negotiations, with the United States (US) situated towards the more intransigent end of the scale and the European Union (EU) towards the more proactive end. While the EU has a more positive attitude towards action on climate change than the US, there are notable divisions among the member countries. Germany, the Netherlands and the Scandinavian states have taken a far more forthright stance on the issue than the countries of Southern Europe and Ireland (OECD 1994). The Netherlands and Germany, for example, have continually tried to impress upon the rest of the EU the importance of introducing a Europe-wide carbon tax, against the resistance of the United Kingdom (UK) and Southern European states (*EC Energy Monthly* February 1996). There is to some degree a split between the countries of Northern and Southern Europe, with those in the North generally being more able to act on the issue of climate change and those in the South, having other policy priorities. States such as Greece, Spain and Ireland have argued that they are entitled to a growth in emissions in order to reach a level of development enjoyed by other EU members. In this context, the issue of 'burden sharing' within a community 'bubble', where Northern members agree to make more sacrifices than Southern members, as part of an overall umbrella target, has been endorsed as a way of reconciling development ambitions and environmental protection objectives. This arrangement is recognised in the Kyoto Protocol. The accession of the states of Eastern and Central Europe to the EU will exacerbate many of these complications (Bergesen et al. 1994) as they are heavily dependent on fossil fuels for economic development and lack the technical capability to make a straightforward transition to a post-greenhouse economy.

The diversity of interests and the disparities in economic and political terms that exist within Europe, have led some to note that the extent to which the EU is able to reconcile these differences will set important precedents for the ability of the rest of the international community to manage the same concerns on a global scale (Bergesen et al. 1994). 'European climate policy is in serious trouble' (ibid.:vii), however, with the apparent failure of the central tenets of its greenhouse strategy: the carbon tax, SAVE (Specific Actions for Vigorous Energy Efficiency), ALTENER (a programme for the promotion of renewable energy sources) and THERMIE (an energy technology support programme).[12] Nevertheless the EU remains one of the more proactive groups from the Organisation for Economic Cooperation and Development (OECD) in the negotiations and pushed (unsuccessfully) for a 15 percent cut in CO_2 emissions by 2010 to be included in the Kyoto agreement.

---

[12] See Collier and Löfstedt (1997) and Grant, Mathews and Newell (forthcoming) for details of these schemes.

The United States

Many of the issues that divide the various groups of states in the negotiations centre on the extent to which legally binding targets for emission reductions are required. The resistance of the US to the notion that the Framework Convention on Climate Change (FCCC) should contain emission reduction targets is widely thought to account for the fact that the convention only calls on parties to 'aim' at stabilisation rather than accept a legally binding target to meet the goal (Brenton 1994). Many countries were so anxious to bring the US on side as a party to the convention that they were willing to drop demands that the treaty contain legally binding obligations to reduce emissions. The EU is singled out as being particularly guilty in this respect (ibid.)

The general line of the US has been to adopt a 'no-regrets' position (Nitze 1994), whereby only action which has benefits for problems other than the greenhouse effect alone is considered (on the basis that there will be 'no regrets' if the threat of global warming does not turn out to be as serious as anticipated). A number of factors explain the intransigence of the US, including the power of the fossil fuel lobbies (Levy 1997; Levy and Egan 1998; Newell and Paterson 1998) and a profligate energy-use culture (Grubb 1991; Paterson and Grubb 1992). The position adopted by the US government has been subject to changes of administration. The Clinton administration has taken a more proactive line on the issue of climate change than the Bush administration.[13] President Clinton, on coming to office in 1992, declared that he would agree to a binding target to stabilise CO_2 emissions at the 1990 levels by the year 2000. This was a commitment President Bush had never been willing to make. Nevertheless the pendulum then swung back towards a more conservative orientation on environmental policy, with the Republican-controlled Congress proposing to reduce or eliminate a range of programmes described as 'vital to implementing the US Climate Change Action Plan' (Anderson 1995:37). At the COP2 meeting the US acknowledged the need for QELROs (quantifiable emission limitation and reduction objectives), but it was also clear about the need to negotiate a political settlement that the Republican-dominated Senate would be willing to ratify. This key (constitutional) dynamic of US climate politics was even more apparent in the run-up to the COP3 meeting in Kyoto, when the Senate passed a resolution, in advance of the meeting, declaring that it would not ratify a Kyoto Protocol that did not contain commitments for less developed countries (LDCs). The tug of war between the government and Congress over this issue continues, with the US delegation declaring at COP4 in Buenos Aires (November 1998) that they were ready to sign the Kyoto Protocol before binding commitments for LDCs had been agreed. The pace of ratification will nevertheless be slow because of the combined force of the Republicans and the fossil fuel lobbies. It is unlikely that the protocol will be

---

[13] Nevertheless Clinton has been frustrated in his attempts to deliver on the climate issue by the resistance that has been mobilised against his proposals, for instance, for a BTU (British thermal unit) tax. Paterson (1993a); Rowlands (1995).

sent to the Senate for ratification before the presidential and congressional elections in November 2000. The goal of opponents to the protocol within the US will be to ensure that action is delayed until after 2004, by which time the task of adjustment to meet the target will be insurmountable.

The importance of the US position on the issue derives from its contribution to the problem (Lee 1995; Rayner 1993), the sizeable contribution it will make to financial mechanisms to implement the convention (Grubb 1991) and its symbolic importance as one of the world's most powerful states. The concern to bring the US on side may also be explained by concern about the loss of competitiveness that might result from 'going it alone' without the participation of all the leading economic powers; a concern that has featured in EU discussions on the carbon tax.

The North–South dimension

One of the most problematic and defining features of the negotiations on the issue of climate change thus far has been the question of responsibility for causing the problem, and following from this, the allocation of burden for responding to the issue. In this context North–South debates have featured highly, where the North has been keen to emphasise that whilst it has contributed most to the problem historically, the future emissions of Southern countries will counter the global effectiveness of any action the North takes to offset climate change (Hyder 1992; Ramakrishna 1990). Many Southern countries have taken the line that the main responsibility for the problem lies with the Northern OECD states and that if these countries are serious about alleviating the problem, they should be willing to fund resource transfers and other forms of financial aid to the South to draw them into a cooperative arrangement. The perception of some countries in the South is that they are being held responsible for a problem they did not cause, and attempts to globalise responses to the problem constitute a strategy by the North to deflect blame from its principal role in creating the problem.[14]

Thus far, many of these tensions have been skirted over via the use of rhetoric in the convention, epitomised by the notion of 'common but differentiated responsibility'; a way of simultaneously emphasising both the need for global cooperation on the issue and the principal part played by the North in perpetuating the problem to date. Parties now have differentiated commitments embodied in the Kyoto Protocol (Table 1.1). The 'Pandora's box' of special pleading for concessions on targets has been opened and the bargaining has begun. And whilst LDCs have not been called upon to accept legally binding reduction targets (a fact that has allowed the US Congress to state it will not ratify the protocol), there has been some movement since Kyoto (see below). The (ongoing) need to have the US as a signatory to the protocol also means that the issue of commitments for LDCs will be an enduring feature of international climate politics, particularly as the emissions

---

[14] Interview with Akumu, 17 July 1996, Geneva.

Table 1.1   *Kyoto targets, key countries over the period 2008–12*

	Target
Australia	+8
Britain	−8
Canada	−6
EU as a whole	−8
Germany	−8
Japan	−6
Norway	+1
Russia	0
US	−7

Note: Cuts in the three most important gases (CO_2, NH_4, NO_2) will be calculated against a base year (1990). Cuts in long-lived industrial gases can be measured against either a 1990 or 1995 baseline.

produced by some G77 countries, particularly newly industrialised countries, rise to G7 levels.

Divisions within the South

The traditional North–South dichotomy increasingly obscures as much as it reveals about the main cleavages in the climate negotiations. It is more helpful to think in terms of there being five groupings within the G77 (least developed country) umbrella. Firstly, there are the AOSIS states, whose position has already been briefly outlined. Secondly, there are the OPEC states, whose position was also discussed above. Thirdly, there is an emerging consensus among China, India, and Brazil and other newly industrialised countries, which as a group have the potential to contribute significantly to future emission levels. These states are being increasingly called upon to accept targets, but are resisting these in the absence of stronger commitments by the OECD states and the promise of technology transfers and climate aid. China, India, Brazil and many other developing states sided with the AOSIS states at Berlin (COP1) in calling for the industrialised countries (Annex 1 states) to accept an obligation to reduce their 1990 CO_2 emission levels by 20 percent by the year 2005 (Grubb 1995), and subsequently to commit themselves to significant reductions in the Kyoto agreement.

Fourthly, there are the majority of states in the G77 bloc, which for the most part, particularly during the earlier stages of the negotiations, have been keen to emphasise the developmental rather than the environmental side of the global warming equation. Since then, the common position for most of these states has been to press upon Northern states both their comparatively small contribution to the problem and their unpreparedness at this stage to accept further commitments under the convention. The consensus on the part of this group, remains that action should only be undertaken by Annex 1 states in the first instance.

The final grouping, which emerged at Kyoto and Buenos Aires, is a small number of developing countries that are willing to pursue voluntary commitments. Argentina dramatically broke ranks with other LDCs by declaring (as host nation to COP4) its willingness to make a voluntary commitment to reduce its CO_2 emissions. Kazakhstan made a similar pledge. The attempt by Argentina to insert a clause in the Kyoto Protocol permitting non-Annex 1 parties to make voluntary commitments was sacrificed at the altar of maintaining G77 solidarity. It was rejected by the G77 countries to avoid a situation in which Annex 1 parties could play off non-Annex 1 parties' commitments against one another (Grubb et al. 1999).

The positions adopted by the different camps within this broad coalition are in a state of continual flux. The COP4 meeting exposed divisions between a number of Latin American countries that are very receptive to the CDM (clean development mechanism) and JI (joint implementation) Kyoto instruments, African states that are pressing for financial assistance for adaptation and capacity building, and India, which is pushing for universal application of the equity principle. What unites Africa, India and China with regard the new Kyoto instruments is their concern for principles to be in place before trading takes place.

The fragmentation of these once fairly robust negotiating coalitions may suggest a potential for change, the development of new alliances or even the formation of a 'winning coalition' (Sebenius 1994). At COP4, for instance, there was a notable increase in cooperation between EU countries and the G77 in the form of formal and informal meetings and the formulation of common positions.

Policy options after Rio

The debate on appropriate policy instruments to address climate change has, unsurprisingly, given rise to many other conflicts around issues of equity, efficiency and effectiveness in the development of response strategies. The post-Rio landscape has been particularly notable for the salience of JI as a proposed method by which to bypass the North–South stalemate described above. The scheme permits states and companies to invest in greenhouse gas abatement projects in states that cannot afford to undertake such projects themselves; states where it is also cheaper to reduce net emissions than in the investors' own countries.

The scheme has been popular among a number of principally OECD states. It has attracted the support of Japan and, most prominently, Norway, which was one of the earliest backers of the proposal. However the scheme has come under fire from a number of quarters, especially environmental NGOs, on the ground that it offers states that are unwilling to change their domestic policies the chance to earn credit for action elsewhere, where the need for emissions reductions is less pressing (Climate Network Europe 1994). Issues relating to the comparability of different schemes, the means by which credit will be agreed (the exact division of credit between participating parties) and how to address the speculation inherent in funding a project that *would not otherwise have happened* in the recipient country, all await resolution.[15] JI is officially endorsed by the Kyoto Protocol as playing a central and legitimate role in parties' abatement strategies (Newell 1998). This will allow governments to sanction industries to participate in projects that earn 'emission reduction units' with other Annex 1 parties. The protocol also created the CDM (clean development mechanism), which allows Annex 1 parties to achieve part of their required reductions by paying money to finance 'climate-friendly' projects in non-Annex 1 countries. In return, investing countries receive credit against their national reduction targets. Funds paid into the CDM will also be used to finance the adaptation costs of the most vulnerable countries and to cover the CDM's administrative overheads. The US is keen that the CDM begin work as quickly as possible and that projects be incorporated that currently have the pilot status of 'activities implemented jointly' (Forsyth 1998). The World Bank is also arguing that backdating the impacts of existing projects would accelerate investment in the CDM. Many LDCs continue to be suspicious of AIJ projects on the basis that they are a strategy by which Annex 1 parties can avoid their own commitments, and that most projects to date have focused on carbon sequestration rather than the demands of LDCs for technology transfer (ibid.:1). Nevertheless, LDCs' acceptance of the CDM in the Kyoto Protocol amounts to a 'volte-face', according to Grubb et al. (1999), given that many of the objections to JI apply to the CDM. The fact that the CDM provides for more institutional (particularly COP) oversight may account for the different attitude towards the two mechanisms.

Before the CDM can play a full part in the implementation of the protocol, a number of outstanding issues will have to be resolved, such as whether crediting can begin before agreement has been reached on the appropriate scientific baselines by which to measure the impact of projects; whether the CDM will incorporate existing AIJ projects and guidelines; whether sink projects are eligible; and the proportion of Annex 1 emission reductions that can be achieved through the funding of CDM activities.

Among the other proposed strategies to deal with the problem of climate change are carbon taxes, tradable emission-quota schemes and more traditional 'command and control' policy instruments. There has been a great deal of resistance to the imposition of carbon or energy taxes both at the regional level in the EU and in individual member states, illustrating the difficulty of states seeking to go it alone in adopting unilateral measures to reduce their greenhouse gas emissions. Market instruments such as tradable quotas are the favoured solution of the US and the

---

[15] For further details of the legal and economic issues that surround the joint implementation issue, see Kuik, Peters and Schrijver (1994).

United Nations Conference on Trade and Development (UNCTAD) and precedent exists for their use, on a national basis at least in the US, where sulphur dioxide permit trading already takes place. Among the problems involved in further developing tradable permit mechanisms are the establishment of distribution criteria and permits 'hoarding' (Grubb 1990; Lunde 1991a). Nevertheless these are preferred over strictly regulatory mechanisms, which are regarded as insensitive to the diversity of energy needs among parties to the climate regime (Grubb et al. 1992b). Permit trading has also made the transition from idea to policy instrument following endorsement in the Kyoto Protocol and the commencement of work at the COP4 meeting in Buenos Aires on the rules and procedures that will provide the framework for these transactions. One issue that has already slowed progress in this area is what has been dubbed 'hot air trading': the selling (and purchase) of permits for CO_2 that would have been emitted were it not for industrial decline. Russia has emerged as a key player in this debate, advocating the right of countries that have suffered deindustrialisation to be permitted to trade emission entitlements in this way.

Implicit in a number of the strategies outlined above is the need to find ways of financing emission abatement, particularly in countries where basic development is the main priority and/or financial, human and technical resources are lacking. Indeed the involvement of poorer countries in the climate agreements is conditional on them receiving aid and technology transfers. Key problems involving the scale and nature of technology transfers and financial assistance, and the ties that may come with them (Hyder 1992; McCully 1991; Nilsson and Pitt 1994; Norberg-Bohm and Hart 1995), remain unresolved. The issue of 'additionality' was one source of disagreement. The US in particular, as one of the principal financial providers of climate aid, has argued strongly that monies for climate change had to come from existing development/aid budgets. Many LDCs rejected this arrangement and called for new resources to be made available to help them meet their commitments under the convention.

The question of which institution should administer these transactions in the long term has also been a source of contention. Since early on in the climate negotiations the question of whether the Global Environmental Facility (GEF) should oversee the funding of climate change projects has featured highly. Many countries from the South have been concerned about the institution's association with the World Bank, with which they have had a fraught relationship in the past. Other points of contention have been the degree to which systems of voting rights employed by the World Bank will form a precedent for the GEF structure[16] (and therefore the extent to which LDCs will have a say in decisions affecting the location and nature of the projects), and the degree of access that NGOs will be accorded (Gan 1993; Nilsson and Pitt 1994). At the Berlin Conference of the Parties the

---

[16] This issue was partially resolved at a meeting held in Geneva in March 1994 to discuss the GEF's replenishment. It was decided that two voting systems would be adopted in a Governing Council, neither of which accords with the 'one dollar one vote' system operated by the World Bank (Rowlands 1995).

GEF was approved as the continuing *interim* financial mechanism for the convention (Grubb and Anderson 1995).

Kyoto and beyond

The debates in the run-up to the Kyoto meeting centred on the scope of commitments (timescale, sources, sinks) the level (differentiation versus flat-rate cuts) and the formulation of appropriate instruments to bring about flexibility in parties' response strategies. The ongoing issue of the participation of LDCs in reduction commitments also made its way back onto the agenda at the insistence of the US.

After ten days of intense negotiations in Kyoto, the final deal that emerged from the diplomatic wrangling reflected the preferences of the US, with its emphasis on joint implementation, emissions trading, the inclusion of six greenhouse gases (including ozone-related chemicals) and the timetable to be employed. Ministers and other officials from over 160 countries reached a legally binding protocol under which the industrialised countries will reduce their collective emission of greenhouse gases by an average of 5.2 percent. Article 3(1) of the protocol allows for individual and joint implementation of commitments 'with a view to reducing their overall emissions of such gases by at least 5 percent below 1990 levels in the commitment period 2008–2012', calculated as an average over these five years. Parties are expected to have demonstrated progress towards reaching this target by the year 2005, by which time a second commitment period should have been agreed. Gains over and above the first-period commitments can be 'banked' against commitments in the second period.[17]

Japan agreed to reduce its emissions to below their 1990 level by an average of 6 percent, the US by 7 percent and the EU by 8 percent; a huge shift on the part of the EU from its initial target of a 15 percent cut. The EU is permitted to meet this target as a 'bubble', with internal burden sharing between member states. There are targets for twenty-one other industrialised countries to be met between 2008 and 2012. Australia, Iceland and Norway are allowed increases in their emissions, while New Zealand and Russia are merely required not to increase their emissions. The final package was stitched together by a deal brokered between the US and China. The developing countries, at the insistence of China, won the debate on new commitments from non-Annex 1 developing countries, so that they are not required to sign up to reductions of greenhouse gases. In return the US retained in the final document the principle of emission trading, but not the details of its operation.

The protocol will enter into force after it has been ratified by at least six countries that together account for 55 percent of the total 1990 emissions in developed countries. In common with the FCCC, the protocol will be implemented via national reports overseen by teams of experts nominated by the parties. There will be no fines for non-compliance, as Brazil had suggested. The COP is encouraged to

---

[17] The US wanted to include a 'borrowing' arrangement whereby parties could borrow entitlements from the second period to enable them to meet their first-period commitments. The proposal was rejected.

develop an 'indicative list of consequences' that parties will face in the event of non-compliance with the protocol. Where these are binding they have to be adopted as an amendment to the protocol.

Drawing up the rules by which the provisions of the Kyoto Protocol will be implemented will provide the majority of ongoing business for the negotiators. The principles, modalities, rules and guidelines for verification, reporting and accountability for each of the flexible mechanisms need to be worked out, as required by Article 17 of the protocol. This means, for instance, determining what constitutes an emission-reduction unit which sectors should be included, who should be allowed to trade and when and by what monitoring system. The issue of 'supplementarity' continued to divide the EU and US negotiators at Buenos Aires: whether there should be a quantitative or qualitative ceiling on how far protocol commitments can be met through the use of these flexible international instruments as opposed to domestic measures (Villagrasa 1998).[18]

The COP4 meeting in Buenos Aires did produce a plan of action that was meant to provide a clear timetable within which outstanding issues should be decided upon. The plan lists twenty-three decisions that have to be made and the issues that need to be resolved in coming to those decisions. It specifies that the rules of implementation (on the flexible mechanisms) will have to be drawn up in time for COP6 (in all likelihood by late 2000 or early 2001). The figure is a fairly arbitrary one with little indication of the plausibility of resolving all the issues by that date (*Global Market Review* 1998). Given the short time span in which all these important issues will have to be discussed, it is expected that much of the business outlined in the action plan will have to be conducted during high-level informal consultations and not confined to the meetings of the COP or its subsidiary bodies.

Many of the most difficult challenges remain, now that the most obvious 'no regrets' options have been exhausted and LDCs are increasingly expected to take on additional commitments. And despite some countries' rhetoric about getting tough on climate change, emissions, for the most part, continue to rise. Japan, the US and Australia all project an emission growth of more than 20 percent up to 2010, and the IEA (International Energy Agency) suggests that EU emissions will increase by 30 percent from their 1990 levels despite the EU's claim to be on course for stabilisation (Grubb et al. 1999).

For some, the pace of progress has been frustratingly slow and has not kept up with advancements in scientific understanding. For others, the action that has already been taken and the range of institutional mechanisms that are in place, provide testimony of a responsible and precautionary approach to an issue that is characterised by uncertainty in many fundamental ways.

---

[18] The US suggested at COP4 that it should be permitted to meet 85 percent of its commitments through flexible mechanisms. The EU, on the other hand, was pushing for a 50 percent cap on their use.

2

Existing approaches: problems and limitations

This chapter reviews two sets of interrelated literature that are relevant to this book; the rapidly expanding literature on global environmental politics; and, more specifically, the literature on the politics of global warming. Finally it looks at how the concept of power is addressed in these bodies of literature, and outlines alternative notions of the concept that may be useful to a project on non-state actors.

2.1 Existing approaches to explaining global environmental politics

Most analysis of global environmental politics emanating from the discipline of international relations remains firmly rooted in a state-centric paradigm (Paterson 1996a; Saurin 1996; Willetts 1993). In other words, it takes as given the preeminent status of the nation-state as the key point of reference in seeking to account for the way in which issues unfold on the global agenda.[1] The principal theoretical tools are derived from various versions of liberal institutionalism in international relations (Paterson 1996a; Saurin 1996; Smith 1993). The use of existing paradigms is particularly ironic given the repeated emphasis by writers from this discipline upon the way in which environmental issues transform our understanding of world politics (Rowlands 1992; Saurin 1996).[2] For most scholars of international relations, global warming is one of a series of collective action problems following in the trail of issues such as ozone depletion. Ward's (1996:850) depiction of the issue highlights this: 'Global climate change is characterised as a collective-action game played by nations through time. The conditions under which conditional cooperation can occur are explored.' Given such a construction of the environmental problematic – as

---

[1] The reference to states throughout the book is not intended to blur the distinction made by Risse-Kappen (1995:283) between the state as an 'institutionalised structure of governance and the state as an actor, i.e. the national Government'. When states are discussed here, as in most work in international relations, it is the national government that is being referred to.

[2] Rowlands (1992:288) argues that 'the dominant interpretations of international relations have become increasingly unable to deal with new environmental issues in world politics ... more recent developments have effectively challenged the traditional interpretations, concepts and frameworks that we use to study the discipline'.

one of inter-state cooperation and institution building – its capture by liberal institutionalist perspectives was almost inevitable.

The most popular branch of liberal institutionalist thinking, as applied implicitly or explicitly to the study of global environmental politics, is regime theory (Smith 1993), which concerns itself with 'norms, rules, principles and decision-making procedures' (Krasner 1983:2) at the international level, and grew out of a concern that, with the decline in the hegemonic power of the US, international cooperation would be detrimentally affected (Keohane 1984). Attention turned instead to the ways in which international institutions might be able to regulate state behaviour in the absence of the stabilising presence of a hegemon. Applied to global environmental change and the problematic of managing the global commons so as to avoid a 'tragedy of the commons' scenario, regime theory appears to offer a useful analytical grounding for the conceptualisation of such problems (Vogler 1995). Regime theory responds to a number of overlapping concerns that traditionally engage International Relations scholars. These centre on the need to restrain egoistic state behaviour and nurture collective agreement in an anarchic international society, the need to deter free-riding and to address the distributive issues that arise when confronting global environmental problems. The appeal of the 'statist' focus of regime theory can also be explained by the view that the state is the primary institution capable of commanding widespread authority and able to direct the sort of large-scale human activity required to meet ecological challenges (Wapner 1996).

The bias in this literature in favour of inter-state analysis, as opposed to sub- or trans-state, limits what regime theory can hope to explain, however. It diverts attention away from a critical set of variables that this book explores in fuller detail. This neglect is all the more surprising in light of Young's (1989c) acknowledgement that 'it would be a serious mistake to overlook the role of transnational alliances among influential interest groups in developing and maintaining regimes at the international level'. Other prominent regime scholars such as Levy, Keohane and Haas also acknowledge that 'the key policy variable in accounting for policy change [in international institutions] ... is the degree of domestic environmental pressure in major industrialised democracies, *not* the decision-making rules of the relevant international institution' (Levy et al. 1993:14, emphasis added). And despite acknowledging that 'NGOs are often the source of policy innovation at the international level, the instruments of diffusion of international norms and practices and sources of national level information at the international level', Haas et al. concede that 'Our analysis has admittedly been state-centric' (ibid.:420).

Moreover the whole notion of governance presupposes 'the management of complex interdependencies among actors (whether individuals, corporations or interest groups) engaged in iterative decision-making' (Young et al. 1991:6). There have been a number of calls for further study of NGOs (Caldwell 1988; Haufler 1995; Levy et al. 1993), but these have not, on the whole, been followed up. Young and Von Moltke (1994:361–2) argue that 'it is critical to deepen our concerns for the pervasive role of non-state actors as players in the processes of regime creation, as the ultimate subjects of many regulations, and as pressure groups in the implementation and operation of regimes. .... No issue-area constitutes a better laboratory in which to study these developments than international environmental affairs.'

There have also been attempts by the transnationalist school to emphasise their relevance (Keohane and Nye 1972)[3] and references by scholars of the environment to the 'effective challenge to the traditional understanding of International Relations' posed by NGOs (Rowlands 1992:296). The one notable exception to this general neglect is the work of Peter Haas on non-governmental expert communities, which develops a cognitive theory of cooperation and goes far beyond the lip-service attention to non-state actors that characterises much other writing on international cooperation (Caldwell 1990; Carroll 1988). Haas's work is reviewed extensively in Chapter 3.

The need to remedy the deficit in our comprehension of global environmental politics constituted by the lack of critical attention to NGOs may be more acute if it can be shown that NGOs fulfil many of the functions of an international regime, and may actually influence states' policies to a greater degree than an international regime is able to. If NGOs are in a position to perform traditional regime functions, it may be that they can significantly determine the prospects of cooperation.[4] Key issues such as free-riding, iterative decision-making, information exchange and the stabilisation of states' expectations, which are so prevalent in the regime literature, are strongly affected by the participation of NGOs. NGOs may apply pressure upon states to cooperate. They may expose instances of non-compliance with a regime and thereby deter free-riding (Hurrell 1995; Jacobson and Weiss 1995; Spector and Korula 1993). By providing information and conferring legitimacy, NGOs can strengthen the institutional capabilities of international regimes, stabilise states' expectations and reinforce the necessity of iterative decision-making. They help to create a 'shadow of the future' by creating an expectation that short-term and tokenistic responses are inadequate and that 'deviant' (non-conforming) behaviour towards a regime will be greeted with hostile publicity and pressure. None of this is intended to suggest that institutions do not also strengthen the position of NGOs in domestic politics (Levy et al. 1993; Risse-Kappen 1995). Relations are undoubtedly reciprocal. The point is that NGOs may also delimit 'the range of legitimate or admissible behaviour in a specified context of activity' (Rittberger 1995:xii) and therefore may be centrally relevant to the regime project. Hence whilst regime approaches illuminate the dilemmas of cooperation, they are less useful in fully accounting for the details and contents of agreements and the configuration of social and political forces that bring them into being and shape their very nature.

---

[3] Reynolds and McKinlay's (1979) transnationalist framework includes mention of NGOs, but only to illustrate their contribution to the thickening of the complex web of global interactions. They note that whilst NGOs may be able to affect the domestic environment of decision-makers, they are unable to affect states' capabilities or objectives, so 'in no way can they be seen themselves as significant actors' (ibid.:154).

[4] Haufler (1995:94) argues that state-centred regime approaches miss 'the important contributions of non-state actors to the creation and maintenance of regimes'.

The analysis offered in this book goes beyond demonstrating the ways in which NGOs may be the 'servants of state policy' (Haufler 1995:108) and makes the more difficult case that state preferences can be conditioned by NGOs. This has the effect of problematising the regime project, because many of the functions that have thus far been attributed to regimes may be shown to be the product of NGO influences. As Humphreys (1996) illustrates, power-based, interest-based and cognitive theories of regimes would all benefit from NGO analysis. The economic functions that corporate lobbyists perform for the state afford them power in global environmental politics, environmental NGOs can highlight the benefits of regime creation and clarify the costs of regime defection, and epistemic communities provide the cornerstone of cognitive regime theories. What emerges from this project may then provide the foundations of a less state-centred theory of cooperation in global environmental politics; an account in which the *a priori* primacy of the state as the key explanatory variable is questioned.

2.2 Critique of regime approaches[5]

The assumptions that regime theory makes about the primacy of states as actors[6] and the unitary and autonomous nature of their existence, provide few insights into a project dealing with the political significance of non-governmental actors. Indeed the narrow focus of regime approaches has blinkered analytical approaches to the study of global environmental politics and inhibited the development of adequate tools with which to understand the importance of non-governmental actors.

One of the key problems with applying regime theory to the global climate regime is that regime theory pursues generalisable hypotheses that apply across issue areas, whereas the particular political dynamics and problem structures that characterise global warming are such that it differs significantly even from seemingly similar global atmospheric issues such as ozone depletion (Rowlands 1995). The notion that different issues give rise to different sets of political relations is not pursued by writers on international regimes, despite Young's observation that explanations of the success of a regime are 'substantially determined by the nature of the problem a regime is expected to solve' (Young et al. 1991:9).

A further problem is that unitary actor analysis, which prevails in regime approaches, precludes assessment of the way in which state interests may be 'learned' or conditioned by interaction with non-governmental actors. It is assumed instead that interests are given. Ward (1996:871) acknowledges this when he notes that 'Game theory cannot constitute a free-standing explanation because it takes states'

---

[5] The very use of the term regime has come under a great deal of fire from scholars (Stein 1990; Strange 1983) who argue that it has 'no conceptual status' (Stein 1990:26). For these writers the term is misleading in respect of the power it assumes international institutions to have.

[6] Strange (1983:337) has argued that regime theory suffers from being rooted in a 'state-centric paradigm that limits vision of a wider reality'.

preferences, beliefs, and strategic opportunities as givens' and 'does not capture all relevant notions of power.' Focus strictly upon the 'ken of international bureaucracies and diplomatic bargaining' (Strange 1983:338) excludes analysis of the manufacture of agendas and the whole prenegotiation phase of policy-making, in which national positions and interests come to be articulated. The degree to which states are free to define their own agendas is therefore exaggerated.

Regime approaches display an assumed rationality about states' choice of cooperative strategies, where actors possess '*consistent, ordered* preferences, and ... calculate costs and benefits of alternative courses of action in order to maximise their utility in view of these preferences' (Keohane, quoted in Grieco 1988:496, emphasis added).[7] Such conceptualisations factor out assessment of the degree of regime change that may accrue from (re)defined or newly defined interests. In criticising this position Wendt (1992:398) notes that 'Actors do not have a "portfolio" of interests that they carry around independent of social context; instead they define their interests in the process of defining situations.' Moreover the value change and social learning central to regime change, which often occur at the national level in the first instance, might be attributable to the activities and lobbying of non-governmental actors. If it can be shown that non-governmental actors have some influence on the interests and expectations that state actors bring to the process of institutional bargaining in international fora,[8] then an important challenge is posited to the way in which we currently seek to explain policy.

Related to this point is the criticism often made of regime theory that it approves of a separation between domestic and international politics.[9] Paterson (1996a:68) argues that even a preliminary understanding of the politics of climate change exposes the 'poverty of this position'. Such a distinction is particularly unhelpful in seeking to understand the importance of NGOs, which, especially when they form international coalitions, cannot be thought of as either exclusively national or international actors, but as both, acting simultaneously across these levels. In this sense NGO politics 'transcend the level of analysis problem' (Nye and Keohane 1972:380). Putnam's (1988) notion of a two-level game (domestic and international) offers a more refined account of the global dynamics of cooperation, in that it offers the possibility of including domestic level analysis: an approach to understanding both the 'internationalisation of domestic politics' and the 'domesticisation of

---

[7] Wendt (1992:392) notes that 'Neo-realists and neo-liberals may disagree ... but both groups take the self-interested state as the starting point for theory. ... Questions about ... interest formation are therefore not important to students of international relations.'

[8] Young acknowledges a stage of 'pre-negotiation including the emergence of an issue on the active policy agenda and the jockeying for position that occurs in connection with framing the issues' (Young and Osherenko 1993:284). Keohane also concedes, 'I assume a prior context of power, expectations, values and conventions' (Keohane 1982:145), 'a given desire for agreements' (ibid.:153). Both writers fail, however, to explore how these may affect, or help to explain, international cooperation.

[9] Milner (1992) cites two reasons for the neglect of domestic politics in the cooperation literature: the centrality of anarchy as a presupposition of any global explanation, and the use of game theory, with its assumptions about unitary rational actors.

international politics' (Nye and Keohane 1972:376). In Putnam's words, 'It is fruitless to debate whether domestic politics really determine international relations or
the reverse. The answer to that question is clearly "Both, sometimes". The more
interesting questions are "when?" and "how?" ' (Putnam 1988:427).

Regime theory tends to overlook the bureaucratic, intragovernmental constraints
that are imposed on government foreign policy, a fact that limits its explanatory use
in relation to global warming.[10] There is a lack of attention to the way in which the
realisation of foreign policy goals may be frustrated by the presence of domestic
interests (Kydd and Snidal 1995). However, looking at the processes behind the
national positions that states adopt in negotiations helps to illustrate the degree of
manoeuvre that negotiators have in international fora. In this regard, Gupta et al.
(1993:18) argue that 'Only by recognising the ongoing political struggle behind governmental positions on climate and by identifying the societal groups involved is it
possible to analyse the dynamics of the formulation of national positions on climate
change.' Conversely governments need to 'anticipate domestic reactions' (Levy
et al. 1993; Milner 1992:493), given that they have a 'fixed investment in a particular
supporting coalition' (Putman 1988:458). This entails developing a 'transnational
equivalent of bureaucratic politics' (Risse-Kappen 1995:9), which penetrates the
'hard shell' of the state, exploring how international politics enter the domestic
realm and vice versa. Putman's account is criticised by Risse-Kappen (1995) on the
grounds that in his analysis governments are the only actors linking the international
with the domestic level, whereas transnational actors can affect both domestic win-
sets and inter-state negotiations, and create linkages between the levels.[11]

The significant point for the purposes of this analysis is that the freedom of the
state to select cooperative strategies can be bounded by non-governmental actors,
who can help to establish a matrix of constraints and opportunities that may guide
state behaviour. As Risse-Kappen (1995:29) notes, 'if governments do have choices
to respond to international pressures and opportunities, there is no reason for excluding transnational actors from the consideration of agents who might influence such
decisions'. In regime language, the argument is that 'transnational relations ... can
alter the choices open to statesmen [*sic*] and the costs that must be borne for accepting various courses of action', and therefore serve to 'provide different sets of incentives or pay-offs for states' (Nye and Keohane 1972:374–5).

Regime approaches are vulnerable to the similar charge that they underplay the
significance of global forces that lie outside the particular institutional arrangement

---

[10] This criticism can be extended to prisoners' dilemma games (Gowa 1986; Kydd and Snidal 1995). There has
been some attempt by game theory scholars to respond to these criticisms by describing the central decision-
maker as buffeted by a myriad of competing interests (Kydd and Snidal 1995). Where this happens, states are
said to perform a collective utility function.
[11] On this basis Risse-Kappen (1995:300) calls for a three-level game where 'level I represents the realm of
transnational/transgovernmental learning, level II the sphere of intra-governmental as well as inter-state bargaining over the negotiating results on level I, and level III the area of domestic politics'.

under scrutiny. To understand 'a given area of international relations' (Krasner 1983:2) (the stated intention of the regime project), a broader focus than a specific set of institutions needs to be adopted. Many political relationships that fundamentally affect the subject matter of a regime are missed by an analysis constructed around the operation of international institutions in a particular issue area. The location of states in a dynamic global capitalist economy, in which the need to attract investment from multinational companies and to avoid capital flight may severely constrain the range of environmental policy options open to a state, is one example. Reference to these broader ties helps to inform understandings of NGO–state relations. The role of industrial groups is perhaps particularly pertinent in this regard, where states' capital accumulation objectives may be argued to confer significant leverage upon the corporate actors that help to deliver those goals (Newell and Paterson 1998).

Much of this section has focused on some of the ways in which analysis of NGOs might contribute towards conventional explanations of international environmental cooperation; the means by which it may supplement or complement existing accounts from the regime paradigm. It has been argued that a project on NGO influence helps to respond to many of the key questions that regime theory sets for itself. It would be unhelpful to exclude analysis of states and international institutions from a project on NGO influence. In talking about the impact of non-governmental activity upon states, however, it is unhelpful to dichotomise the relationship in such a way that the emphasis is on showing how one actor is more powerful than the other.[12] As Risse-Kappen (1995:13) argues, 'One can subscribe to the proposition that national governments are extremely significant in international relations and still claim that transnational actors crucially affect state interests, policies and inter-state relations ... one does not have to do away with the "state" to establish the influence of transnational relations in world politics.' Power relations are reciprocal. As Elliott (1992:13) points out, 'The relationship between states and non-state actors is a constitutive one. Both states and NGOs may enhance or constrain the actions of the other.' In this sense, what this book is trying to achieve can be located as part of a project begun by Risse-Kappen (1995): of bringing transnational relations back in, where regime approaches have lost sight of them. The specific brief here, though, is to demonstrate ways in which attention to the political impact of NGOs upon the processes of international cooperation (broadly conceived) on climate change offers rewarding sources of explanation. To do this effectively, structural as well as constructivist accounts are brought to bear, taking the emphasis of transnationalism as a starting point but developing it in new directions.

---

[12] For example, Goldmann and Sjöstedt (1979:26) adopt the realist view that international actors only exist if they are 'comparable to states in terms of their power in international matters'.

2.3 Current approaches to the study of global warming

The recognition that global warming is at once a fundamentally political as well as scientific question developed slowly in academic circles from the late 1980s onwards.[13] Thinking about the complexities that hinder attempts to contain the problem in a way that satisfies the disparate demands of all interested parties provided the initial focus (Sebenius 1991; Skolnikoff 1990). Much of the earlier literature on the subject was also understandably concerned with the scale of the threat posed by the issue and the sorts of crisis management packages that could be drawn up to handle the problem on a short-term basis (Boyle and Ardill 1989; Leggett 1990).

The key political debates that are addressed in this literature are: (1) whether short-term action of a precautionary nature is justified on the basis of existing scientific knowledge about climate change (Gray and Rivkin 1991; Leggett 1990, 1991); (2) whether making economic sacrifices now will impose unnecessary costs (Bate and Morris 1994; Beckerman 1992; Beckerman and Malkin 1994; Cline 1993; Hope et al. 1993); and (3) the means by which abatement measures can best be brought about in terms of efficiency, equity[14] and overall effectiveness. This has involved a lively debate on the extent to which traditional command-and-control (regulatory) or market mechanisms can be effectively employed to meet established targets (Beckerman and Pasek 1995; Grubb 1990; Hahn and Stavins 1993; Stavins 1993). Along similar lines, there has been an attempt to explore the possibility of developing a 'winning coalition' of parties that are willing to advance action on the issue ahead of the rest of the international community (Fermann 1997). It is hoped that such a coalition could transcend the stalemate (particularly between North and South) that has characterised other complex negotiations on law of the sea, and other issues (Hampson 1990; Sebenius 1991; Sydnes 1991; Tolba 1989). Lessons have also been drawn from the 'success' of the international response to the issue of ozone depletion in trying to map out the most productive future course for the climate negotiations (Benedick 1991c; Rowlands 1995).

The political issues that global warming presents the international community with feature far more highly in the literature on global warming (Richardson 1992; Skolnikoff 1990; Wirth 1989) than accounts and explanations of the development or overall content of policy responses to date (Arts 1998; Paterson 1996b; Rowlands 1995). There has been a great deal of assessment of the Climate Convention in terms of its meaning, content and likely future course (Grubb 1992; Mintzer and Leonard 1994; Nilsson and Pitt 1994; Paterson 1992). And yet, as Paterson (1996a:59) points out, 'There remains an analytical gap in the understanding of processes of international cooperation with regard global warming.' Many accounts describe the series of events that led to the signing of the convention in Rio in 1992.

---

[13] See Newell (1995c) for a review of texts on the politics of climate change.

[14] There has also been an attempt by some writers to explore in more detail the ethics of climate politics. See for example Grubb (1995b), Paterson (1993b, 1994) and Shue (1992).

They identify future key issues, but do not seek, as such, to develop an explanation for the politics they describe.

Whilst Mintzer and Leonard (1994) offer a valuable insight into the formation of the Climate Convention from the point of view of a range of actors involved in the process, their volume does not lend a great deal to our understanding of the political impact of the non-governmental actors researched here. Faulkner (1994) and Rahman and Roncerel (1994) provide NGO perspectives on the key issues and defining moments of the negotiations, and other writers provide, in passing, comments on the influence of different NGOs (Andresen and Wettestad 1992; Grubb 1995a). In the majority of these accounts, however, the level of analysis is the state, and international institutions the fora in which the key bargains are struck, and hence are the presumed loci of explanations of cooperation. Where mention of NGOs is made, it is brief and largely unconnected to outcomes, at least in an explicit sense. Similarly there has been little attempt to connect the literature on the climate politics of individual countries with wider explanations of regime change. Implicit in the absence of an attempt to connect these literatures is an assumption, already challenged, about the separation of domestic and international politics.

Within IR, as Ward acknowledges, 'The key issue for most of the literature on global climate change is whether international legal instruments and institutions can be constructed within which nations will cooperate', and whether these can 'change the incentive structure in ways that facilitate cooperation.' (Ward 1996:860) The abstract of Ward's article on game theory and climate change captures the essence of the climate change problem, viewed from a traditional IR perspective: 'The model clarifies the bargaining tactics used by nations in the negotiation of the Framework Convention on Climate Change and the reasons why there may be collective action failure. The model also illuminates issues of regime design' (ibid.:850). This framing of the issue reveals much about the approach of traditional IR perspectives. (1) its reduction of the politics of global warming to a focus on inter-state bargaining; (2) its emphasis on regime design (problem solving); (3) its brief to explain the conditions under which cooperation can occur and how 'collective action failure' may be avoided; (4) its assumption that 'nations can be viewed as unitary actors making choices between strategies so as to maximise their expected pay-offs' (ibid.:854); and (5) its neglect, besides passing reference, of agenda formation, domestic politics and NGOs. Ian Rowlands' book *The Politics of Global Atmospheric Change* (1995) explains the politics of climate change from an IR perspective, employing power-based, interest-based and cognitive-based regime approaches. Passing reference is made to the role of interest groups in the emerging climate change regime, but the book explicitly grounds itself in a traditional, institutionalist analysis of the problem.

Paterson's (1996b) work on explaining the politics of climate change is the most comprehensive attempt to date to theorise the history of international cooperation on the issue, and addresses regime perspectives as part of a wider account of the approach that IR theories would adopt to account for the politics that led to the creation of the Climate Change Convention. His work further emphasises the limitations of traditional IR theory as it might be applied to global climate change. The only other work that explicitly integrates an NGO-based analysis into a discussion of the politics of climate change is Arts (1998). Much more is said about this work in Chapter 6. For now, it suffices to note that he restricts his analysis to environmental NGOs and not the broader spectrum of NGO activity, and that his approach differs sharply from that taken in Chapter 6.

The sets of actors that form the basis of this research have not, on the whole, received significant attention in the literature on global warming. The extent to which this is so varies according to the group of actors in question and is reviewed in Chapter 1 in the section 'Why these actors?', and more fully in the Chapters which follow. There remains very little substantive work in the wealth of literature on global warming that directly deals with the issue of how to conceptualise the political influence of NGOs.

2.4   Reconfiguring political influence

In this final section, existing approaches to studying political influence, as it pertains to non-state actors, are discussed in more detail. The work reviewed has been selected on the basis of its perceived relevance to the issues addressed in this book. Therefore the section by no means offers a comprehensive survey of debates around one of the most problematic, and at the same time 'elusive but indispensable' (Wrong 1979:vii), areas of study in the social sciences.[15] Crenson (1971:109) describes the issue of 'how is political influence to be measured?' as 'one of the most frequently posed and difficult problems of social science'. The concept of power is argued here to be far more complex and contested than traditional scholars of IR have led us to appreciate.[16] Understandings of 'power' are often implicit in IR literature. It is taken as given that it is about states and coercive capabilities. It is assumed to be about possession, authority and domination/submission; power as mastery in the classic 'power over' formulation of the concept, evident in a number of the defining texts of the IR discipline (Bull 1977; Morgenthau 1979). In this sense, power has traditionally been thought of as a zero-sum concept in which one state's gain is, by definition, another's loss.

The forms of political power associated with non-state actors are not those which have traditionally concerned IR scholars. The different approaches vary in the extent to which they privilege relative over absolute gains, but the notion that states are the only possessors of power in global politics prevails (Wendt 1992). The language of hegemons (Keohane 1984), 'great powers' or the 'balance of power' is indicative

---

[15]   See Clegg (1989), Lukes (1974) and Wrong (1979), for a broad overview of debates on power.

[16]   For some classic IR formulations of what power means in international politics, see Goldmann and Sjöstedt (1979), Morgenthau (1979) and Waltz (1979). These texts illustrate Goldmann and Sjöstedt's (1979:1) point that 'Power in the sense of who controls whom remains at the core of the study of politics, including international politics.'

of this. The literature on global environmental politics, while not using these security-oriented notions of power, imports from other work in IR assumptions about power in international society, with a bias towards states and institutions. Coming to terms with the political importance of NGOs, however, requires more imaginative thinking about political power. As Elliott (1992:12) argues, 'the idea of power needs to be expanded if we are to understand how non-state actors such as NGOs are influential and important'. Power can be understood as something more than a relationship of domination, subordination, authority and control, where gains in the influence of one group are not necessarily at the expense of the influence of others (Bachrach and Lawler 1981).

It is almost a truism that regime theory lacks a sophisticated understanding of power in global politics (Milner 1992; Strange 1983; Underdal 1995; Young and Osherenko 1993). This indeed has been a point of contention with the regime project for structural thinkers in particular (Strange 1983; Stein 1983, 1990). Yet the neglect of power they are bemoaning is a structural, state-centred notion of power. The absence of critical thinking about power criticised here is the neglect of less state-centred, cognitive and second-dimensional forms of power. Nevertheless the issues of how the 'pay-off matrix' is structured, how options are constrained and actor involvement limited, which all feature in structural critiques of regime approaches, are useful in an NGO account of political influence.

Traditional understandings of power need to be considered alongside NGO-centred dimensions. There is in any case rarely one form of power operating at any one time. Seemingly individual acts of power by particular actors are always embedded in wider configurations of power and influence that make the individual exercise of power significant and possible. This is overlooked by Young, who notes that analysis of regime transformation is weakened by 'the fact that we continue to lack a satisfactory measure of power despite numerous efforts to devise a useable *metric* or *index* in this area' (Young 1989c:98, emphasis added). Reducing the concept of power to institutional variables excludes consideration of the numerous sites of power in global environmental politics. To redress the absence of critical attention to power in international environmental politics, particularly as it pertains to non-state actors, ideas about power from further afield than IR and environmental politics are drawn upon in this book to make sense of the dynamics of the global warming issue.

Providing an anatomy of forms of non-state power in global climate politics involves a number of methodological problems. The most obvious is how to attribute influence to particular players or courses of action, to tease out which actor or incident is responsible for which outcome; a reconfiguration of the age-old dilemma 'what is cause and what is effect?' (Risse-Kappen 1995:288). As Young and Osherenko (1993:243) note, 'The methodological problems involved in separating out the contributing factors and assigning weights to each in an effort to explain or predict regime formation are formidable.' For this reason that the analysis here seeks to go beyond 'tracing institutional features back to some actors' preferences' (Mayer, Rittberger and Zürn 1995:419) in a direct and linear manner. The focus is not causal hypotheses about behaviour, but the ways in which NGO actors may be influential in both direct and tacit ways. It is about accounting for both perceived and observable influences, rather than measuring or quantifying linear and causal propositions in a deterministic fashion.

In this sense familiar A–B formulations of the power of one actor over another, which assume there are no intervening variables, other actors or complicating issues are unhelpful. They may be more appropriate for attempting to assess comparatively which is the more powerful – the NGO actor or the state – than for exploring the ways in which NGO actors may be able to influence state behaviour. The core issue here is how to gauge influence. One definitional issue is the potential difference between power and influence. Bachrach and Lawler (1981:25) argue that there exists a circular relationship between the two concepts, where 'Power implies its use, use implies influence, and influence, by definition, implies power.' They distinguish between power and influence on the grounds that 'Power is inherently coercive and implies involuntary submission whereas influence is persuasive and implies voluntary submission' (ibid.:12). For Arts (1998), in his work on NGOs and climate change, power is equated with a more or less permanent ability to affect policy outcomes, while influence is more fleeting, occasional and episodic. Cox and Jacobsen (1973:3) distinguish between power as capability (the aggregate of political resources available to an actor), and influence (the manifestation and exercise of capability). The two terms are taken throughout this book to be intimately related, to the extent that they can be used interchangeably without need for qualification. Nevertheless, the discussion in each chapter of the structural factors and bargaining assets of each actor explores this interrelationship where bargaining assets fit more closely with the way in which these writers define influence and structural factors resonate with their notion of power.

Evidence of influence can be found in the degree to which other actors take notice of NGOs, in the documentation of successful campaigns in the context of observable change, but rests also on a counterfactual argument about whether the outcomes could be expected in the absence of NGO campaigning (Arts 1996; Elliott 1992).[17] The issue here is not power as a 'quantitative concept' (Dahl 1963:336; Russell 1946:35), as Russell contends, whereby the quantity of achieved intended effects can be gauged. It is the extent to which the types of influence that particular groups bring to bear help to us to understand global climate politics. Firstly, it is very difficult to quantify influence in any meaningful or analytically helpful way, and secondly, such an assessment would be oblivious to the difficulty of actors achieving their goals. As Emmet queries, 'is it useful to measure power by the *number* of achieved effects unless we take into consideration the *kind* of effect? (quoted in

---

[17] It is counterfactual speculation about whether an outcome may have been different under a different set of circumstances that has been the focus of most behaviouralist criticism. Lukes (1974:39) argues in defence of the approach that 'It does not follow that, just because it is difficult or even impossible to show that power has been exercised in a given situation, we can conclude that it has not.'

Clegg 1989:72, emphasis added). A project of this nature has to be sensitive to the fact that actors start from different points.

The forms of power considered here are the power to assert frames of interpretation and meaning upon problems, and the power to create expectations and organise preferences. This includes the power to determine what is and what is not policy-relevant knowledge, to inform cognitions and to confer legitimacy upon particular courses of action. The notion that organisation of forms of knowledge constitutes a form of power is very Foucauldian in emphasis (Rouse 1994) and relates to his notion of 'regimes of truth' (Foucault 1980:131), where particular mechanisms, techniques and procedures of power are accorded value and status. For example the status accorded to scientific knowledge informs understandings of the policy influence of the Intergovernmental Panel on Climate Change (IPCC). Young indicates that these forms of power have a role to play in understanding international regimes: when talking about epistemic communities he notes the importance of 'identifying and framing the issues at stake in shaping the way participants understand their interests' (Young and Osherenko 1993:249). It is important, then, to look at the processes by which particular issues are factored into policy debates, and therefore also, by implication, the means by which other issues, interpretations and accounts are sidelined or never acquire salience.

Keeley's (1990) application of a Foucauldian analysis to regimes offers an interesting precedent for exploring the relationship between power and knowledge. Keeley's use of a social constructivist[18] approach is useful in exploring 'subjugated knowledges' (ibid.:91), for as Patton (1979b:115) notes, 'a discourse which is accepted or presents itself as the truth always conveys certain effects of power, notably the exclusion of other kinds of discourse'. Regime approaches 'lose a full sense of the world as contestable and contested' (Keeley 1990:84) by ignoring the process of prenegotiation, where issues are marginalised or privileged by the way in which interests and positions are negotiated and determined. In a complex issue area such as global warming, 'the configuring and weighting' of the different aspects of the issue are 'part of the politics of the "negotiated collective situation"' (ibid.:95). Such an approach to power challenges conventional IR notions of a 'pre-social' (ibid.:99) state of affairs prior to institutional interaction (further exposing the fragility of the domestic–international divide). It also enables us to look at how a range of actors are involved in the process of shaping interpretations and problem definitions, and therefore provides a more inclusive formulation of power in world politics.

The need to cast the analytical net wider than the state is emphasised by Patton (1979b:125) who argues that 'One impoverishes the question of power when one poses it uniquely in terms of legislation, or of the constitution, or only in terms of the state or state apparatus. Power is much more complicated, more dense and

---

[18] For Keeley, behaviour only 'makes sense … within the framework of a construction of reality, which may well affect as well as reflect networks of relations in a society' (Keeley 1990:90–1). See also Hannigan (1995).

diffused than a set of laws or an apparatus of the state.' State-centred conceptions of power limit the kinds of question that can be asked about power and reduce the problem to one of force, law and the power of institutions.

Making connections between some of these ideas and the operation of power in global climate politics is the task in hand. In seeking to do this, notions of non-decision-making (Bachrach and Baratz 1967), anticipated reaction (Crenson 1971; Freidrich 1937)[19] and other ideas that address less observable influences are drawn upon to show how interests can be internalised and anticipated. These ideas come under the broad heading of second-dimensional power (Lukes 1974). The second dimension of power refers to the way in which issues are kept off the agenda and the means by which certain groups are advantaged by political institutions. These advantages 'develop over time as a result of a whole range of acts, intended and unintended consequences, and the repetition of routines' (O'Riordan and Jordan 1996:78). Non-decision-making can refer to the influence of routines, close ties between groups and the reputation of actors, which conspire to keep grievances unarticulated and *potential* issues from realisation. These are what Kripps (1990:176) refers to as 'non-active' forms of power, and enable us to get beyond notions of power based on intent. As Clegg (1989:77) notes, 'Power can produce a situation in which there is little or no behaviourally admissible evidence of power being exercised, but in which, nonetheless, power is pervasively present.'

Looking only at the initiation, opposition, vetoing or altering of proposals in formal meetings, or at what might be termed 'act-response sequences' (Wrong 1979:8), as regime scholars do, misses issues such as the mobilisation of bias and the operation of anticipated reactions (Clegg 1989; Friedrich 1937). Where perceptions of power are themselves relations of power, and where inaction may also be evidence of the exercise of power, Rouse (1994:106) notes, 'actors may exercise power unbeknownst to themselves ... if other agents orient their actions in response to what the first agents do'.[20] The forms of power implied by these notions are manifested by the power to ensure that things do not get done, a power, based not on outward and observable interventions, but 'exercised by confining the scope of decision-making to relatively "safe" issues' (Bachrach and Baratz 1962:948).[21] Bacharch and Baratz's work is especially useful in looking at forms of power that may be

---

[19] Crenson's work on air pollution policy in US local politics revealed that, 'Though the corporation seldom intervened directly in the deliberations of the town's air pollution policy-makers, it was nevertheless able to affect their scope and direction' (Crenson 1971:107).

[20] Lukes (1974) also argues that actors can be unaware of how their actions are interpreted and of the consequences of that action, yet still be exercising power.

[21] The most explicit articulation of Bachrach and Baratz's position on the second dimension of power is where 'A devotes his energies to creating or reinforcing social and political values and institutional practices that limit the scope of the political process to public consideration of only those issues which are comparatively innocuous to A' (Bachrach and Baratz 1962:948).

attributable to actors that are not strictly part of the decision-making circle, such as the mass media.[22] But it is also useful in developing the idea that the size and importance of particular actors to governments, most obviously corporations, means that they cannot remain outside the realm of policy, even if they take no active part in actual deliberations.

Participation, as Crenson's work (1971) bears out, does not provide a complete guide to influence. 'The mere reputation for power, unsupported by acts of power', can be sufficient to restrict the scope of decision-making and lead to situations of 'politically enforced neglect' (Crenson 1971:177, 184). To argue against this is to imply that governments do not assume the interests of actors on their behalf, that governments do not perceive power.

This can be extended to areas of policy that are not often considered when talking about environmental policy. The inclination of regime writers to focus on the narrow domain of institutional arrangements means that they tend to overlook the importance of what is referred to in this book as 'cross-issue influence' (see Chapter 5). Cross-issue influence refers to the direct and tacit power of organised industry in areas of policy-making other than the environment, which nonetheless condition the available negotiating space in relation to climate change. In other words, looking in isolation at the formal arena of policy-making on global warming would factor out potentially very important pressures and influences from other areas of policy (such as energy and industry) that affect a government's policy position on climate change. Perceptions of influence are not contained within one issue area.

Equally, it is possible to make a case that non-state actors also tailor their impressions about which policy proposals to press upon governments to an expectation of what a particular government will be receptive to. In this sense also, a notion of anticipated reaction can be said to delimit the sorts of influence that non-state actors attempt to press upon governments. Crenson (1971), for example, shows how the activists in his case study tended to frame their demands in ways that they anticipated would achieve a less hostile political reception, and on this basis omitted some requests and adopted others accordingly. Liberal institutionalist accounts in IR overlook these prior processes of contestation; the politics that precede the conferring of status upon an issue by an institution, and the ways in which potential issues are filtered. This is not to say that the power to enforce inaction necessarily lies outside the realm of visible political action, as it clearly operates within institutions. But where it potentially does, a focus beyond the confines of strictly institution-based analysis is justified. If these tacit forms of power can be attributed to NGO actors, then we may be better placed to account for why there are 'seemingly important decisions that are never made, significant policies that are never formulated, and issues that never arise' (ibid.:4).

---

[22] For Bachrach and Baratz, grievances outside the formal arena of decision-making have to have been articulated and then neglected (the second dimension of power), whereas for Lukes (1974) the third dimension of power explores how some issues are subjugated within a system and therefore remain latent and unarticulated concerns.

The key issue surrounding the use of second-dimensional power is whether there has to be evidence of clear intent on the part of those actors said to be exercising power. For Wrong (1979), the potential to exercise power is inherent in the meaning of the word and unforeseen effects can accompany deliberate exercises of power, but their importance is only meaningful when they act as by-products of *intended* influence.[23] Dowding (1991:138) develops this point by arguing that 'if you get what you want without trying, because of the law of anticipated reactions, then you are lucky. If, however, this operates effectively as a result of the reputation one has managed to produce for oneself ... then one has produced reactions through an exercise of power.' For these writers, unintended effects have to be part of the exercised power of actors (Dowding 1991; Goldmann and Sjöstedt 1979; Wrong 1979).[24] Wrong (1979:5) argues in favour of 'carefully distinguishing between outcomes that are intended and those that are not'. It is possible to argue that this is a virtually impossible and not necessarily helpful exercise, particularly for the project here, where it is the perceptions of governments that are important and not necessarily the true degree of power that is ascribed to an actor. Therefore whether or not an outcome is intended is not especially significant. In his analysis of the political influence of NGOs, Arts heavily emphasises the importance of intentional intervention in the political arena, but interestingly combines this with a counterfactual implication 'that one's policy goals would not have been achieved or to a lesser extent if one had not intervened' (Arts 1998:301).

Another criticism of the utility of second-dimension conceptions of power is Dowding's (1991:114) argument that to assert the presence of these forms of less observable power one has to look at how those actors, for whom influence is claimed, react to the mobilisation of contrary interests. Here the issue is whether conflict has to exist in order for us to be able to say that there is a relation of power. For some writers, for example, an actor is not powerful unless its interests come into conflict with those of other actors and are *seen* to win the confrontation (Bachrach and Lawler 1981; Polsby 1980; Weber 1947).[25] For others the lack of conflict between actors may be evidence of a tacitly similar agenda and a covert consensus that action is undesirable. Lukes, for example, argues that power can be used to 'avert or preempt' conflict and can be 'cooperatively held and exercised' (Lukes 1994:2). In this way, cooperation also provides evidence of power relations. Many of the conceptual

---

[23] On the question of explicit articulation of interests, Wrong (1979) argues that it is sufficient for A not to deny a reputation of power or to allow his [sic] own interests and goals to be furthered as a result of his reputation. The problem with this line of argument arises when an actor may deny a reputation for being influential (for example, a corporate lobby group may do so in order not to be seen to be acting against the public interest). This would surely not mean that as far as policy-makers are concerned that group is no longer influential, in which case its perceived influence remains important.

[24] For Lukes (1994) the issue is not the actual *production* of such effects, but the *capacity* to produce them that is significant.

[25] Polsby (1980) argues that identifying the actors that prevail in decision-making is the best way to determine which groups have most power, on the basis that direct conflict between actors presents a situation most closely approximating a test of their capacity to affect outcomes.

tools outlined above, and used in this book, clearly incorporate this latter view, that evidence of conflict is not the only way of knowing whether power is being exercised.

There are few precedents for examining the nature of less visible political influences in IR. Writing about the general neglect of these dimensions of power in the study of politics, Crenson (1971) offers an explanation that is also useful in understanding the neglect of this approach in IR. He argues that behaviouralist research traditions have served to focus attention on actions, decisions and institutions. In other words, issues of neglect and non-issues have not been considered worthy of attention. Whilst pluralist theory does acknowledge the existence of indirect influence (Dahl 1957, 1961), it does not make much of it (Crenson 1971).[26] Crenson attributes this to a misguided concern for objectivity and to the view that there can be no explanation for the existence of non-issues. Areas of political neglect are assumed to be explainable by a lack of political interest. Dahl (1961:93) argues, for example, that 'any dissatisfied group will find spokesmen [*sic*] in the political spectrum'.

Although not developed or extensively applied either in the regime literature more generally or in the literature on global environmental politics, there is implicit precedent for the use of less visible power articulations in regime explanations. Keohane's classic definition in *After Hegemony* of cooperation occurring 'when actors adjust their behaviour to the actual *or anticipated preferences* of others through a process of policy coordination' (Keohane 1984:51–2, emphasis added), and Milner's reference to the way in which cooperation is about 'adjusting ... policies *in anticipation*' (Milner 1992:468, emphasis added) are evidence of such precedents. Strange is more explicit in referring to the 'vast area of non-regimes', where 'highlighted issues are sometimes less important than those in shadow' (Strange 1983:338).

Moreover the prisoners' dilemma (Axelrod 1990), which informs so much thinking on cooperation, is grounded in a notion of tacit understanding and the unarticulated. It is premised on the idea that actors anticipate the reactions of other actors. On the issue of reputational power too, Axelrod argues that 'knowing people's *reputations* allows you to know something about what strategy they use even before you have to make your first choice' (ibid.:151, emphasis added). Nevertheless, these uses of power are often not acknowledged as second dimensional. As noted above, they have also not been applied in global environmental politics generally, or in relation to non-state actors specifically.

Combinations of the analytical tools described above will be used to account for the political influence of the non-state actors that are the subject of this book.

---

[26] Dahl (1961:164) argues, for instance, that political leaders 'keep the real or imagined preferences of constituents constantly in mind in deciding what policies to adopt or reject'. This would appear to be a concession not only to the idea that indirect influence exists, but also to the notion that interests can be perceived and internalised by decision-makers outside formal decision-making arenas. The classic pluralist formulation of power, where 'A gets B to do something he *would not otherwise* do' is of itself inherently speculative, and therefore prey to all the pluralist criticisms of second- and third-dimensional power.

3

Knowledge, frames and the scientific community

3.1 Introduction

We should admit ... that power and knowledge directly imply one another; that there is no
power relation without the correlative constitution of a field of knowledge, nor any knowledge
that does not at the same time presuppose and constitute power relations. (Foucault 1977:27)

Traditionally, scientific advice has been assumed to have considerable importance in
assisting states in their attempts to come to terms with environmental problems characterised by complexity and uncertainty. This working assumption is embodied in
international regime arrangements that institutionalise scientific input, and is the
basis of widespread assumptions in writing on global environmental politics that
greater scientific knowledge, or the existence of a consensus among a scientific community, enhances the likelihood of political cooperation (P. Haas 1990a). The same
is thought to be true in the case of global warming (Skolnikoff 1990). It is notable in
this respect that Article 9 of the Framework Convention on Climate Change, on
the Subsidiary Body for Scientific and Technical Advice (SBSTA), anticipates the
need to keep the future negotiations well informed by 'assessments of the state of
scientific knowledge relating to climate change' (UNFCCC 1992: Article 9).

3.2 Structural factors/bargaining assets

Scientific communities gain access to global policy formation on a number of
grounds. Firstly, the perception that scientists are able to reduce uncertainty and
therefore reduce unnecessary political risk by helping states to define their interests,
entitles scientists privileged access to decision-makers. Secondly, creating an expectation that knowledge will become more consensual and offer clearer guidance as
research develops, provides the scientific community with an influential base in
terms of assuring an ongoing input in the policy process. Thirdly, the perception
that scientific advisers can identify and describe problems, account for processes
and define realistic options for societal response (Wynne 1994) enables them to
have a significant formative voice in the debate. Finally, the perception of science
as a non-political voice (Moss 1995) amid competing political voices affords
scientists, as the relayers of that knowledge, a privileged status in international
political negotiations. The advice scientists offer is thought to be 'neutral', and

above the political fray. For example the impression that Working Group 1 is 'detached from political bias' is an important factor in its success, according to Jaeger and O'Riordan (1996:16). This understanding of the role of science explains why politicians depend on the scientific community to confer upon policy responses a degree of legitimacy by lending their approval. By seeking to validate arguments by reference to science, policy-makers allow the scope of future options to be partly constrained by the evolution of scientific research and ensure an ongoing demand for scientific advice.

These functions of scientific knowledge (which translate into influential assets for those who represent that knowledge) are useful in understanding why the discourse on appropriate responses to climate change has been a scientific discourse, with claim followed by counter-claim about the validity of climate change projections, so that science continually provides the reference point for action. Liberatore (1994:190) notes that ever since global warming reached the international agenda, 'scientific findings and uncertainties are used as arguments for deciding between different courses of action'.

This chapter is concerned with the political influence that the expert and knowledge-based community constituted by Working Group 1 (WG1) of the Intergovernmental Panel on Climate Change (IPCC) has upon international policy on global warming. Such knowledge-based communities have been referred to as 'epistemic communities'[1] in a literature by Peter Haas, which explores the role that experts have been able to play in political efforts to confront a range of environmental (and other) issues characterised by uncertainty. Epistemic communities are defined as 'knowledge-based groups of experts and specialists who share common beliefs about cause and effect relationships in the world and some political values concerning the ends to which policies should be addressed' (P. Haas 1990a:xviii). Characteristics of the epistemic community are a claim to authority within their profession, independence (from government), cohesion (in terms of a consensual knowledge base and a common interpretive framework), international links and a common set of political objectives grounded in a shared normative framework.

Haas's notion of epistemic communities is relevant for understanding the influence of WG1 in the global warming debate for a number of reasons. Assessment of the role that an epistemic community can perform in the evolution of policy responses to environmental threats, concerns itself with the political impact of a group of non-governmental actors upon global environmental policy, and is therefore directly relevant to this project. Haas states quite categorically, that 'International environmental cooperation is generated by the influence wielded by specialists with common beliefs' (Haas 1990a:xxii).

---

[1] For a more general account of the importance and uses of an epistemic community analysis in IR see Adler and Haas (1992). The epistemic community model has been applied to case studies as diverse as nuclear arms control, management of whaling and efforts to mitigate stratospheric ozone depletion (Haas 1992:5). For a variety of definitions of what is constituted by an epistemic community see E. Haas (1990:40–2) and Paterson (1996b), specifically in relation to global warming.

A factor that further suggests the usefulness of Haas's ideas to explaining global climate politics, is the degree of scientific uncertainty that surrounds understandings of climate change; a factor that should enhance the influence of the community, according to the hypothesis. The scale of the involvement of the scientific community with the issue also suggests the importance of scientists to the climate policy debate. As Litfin (1994:192) notes, the 'level of political involvement by scientists in the climate change issue is unprecedented in international politics'.

Haas (1990b:359) refers to an 'emergent form of an epistemic community' in relation to global warming, an 'incipient' community at best, and suggests that the applicability of the hypothesis to global warming 'is much less clear' (ibid.:358).[2] However, given the reasons outlined above, Haas's approach will be used selectively to understand the political influence of WG1. It is used alongside other relevant work, such as that by Karen Litfin (1993, 1994, 1995), which takes a more discursive approach to understanding how the influence of 'knowledge brokers' operates. The account of WG1's influence provided in this chapter emphasises, more strongly than Haas's work, the political nature of the knowledge-selection and production process, which belies the communication channels between expert and policy-making communities. It also pays more attention to the two-way interactions between the scientific community and policy-makers, than the more linear one-way transfer of knowledge/power from expert to policy community implied by Haas's work. This is central to understanding the nature of WG1's influence.

Given the issue of bureaucratic entrenchment in particular, where scientists become government scientists, and the intergovernmental nature of the IPCC panel more generally, applying the term 'non-governmental' to the scientists involved with the IPCC may be stretching the meaning of the term. The inclusion of 'intergovernmental' in the title of the panel is thought by Lunde (1991c) to denote evidence of an explicitly political framework. For many, the creation of the IPCC constituted an attempt by governments to assert control over the climate change agenda (Boehmer-Christiansen 1994a, 1994b; Vogler 1995).[3] The WMO (World Meteorological Organisation) invited governments to join the IPCC by appointing their own WMO representatives, so that 'countries can effectively exercise a veto on any individual' (Boehmer-Christiansen and Skea 1994:4).[4] Moreover the IPCC is heavily dominated by scientists who have either worked for government-funded laboratories, or are dependent upon government contracts (ibid.). That said, given that other groups looked at in this book also work very closely with governments, often appear on their negotiating teams and work as policy advisors, it is clear that the line between what constitutes governmental

---

[2] Susskind (1994:75) claims that in relation to global warming there is a 'fragmented epistemic community'. For Lunde, on the other hand, 'it is not difficult to find approximations to epistemic communities active in greenhouse consensus-building' (Lunde 1991c:147).
[3] The interesting assumption here is that scientists had been permitted to set the pace before that time as part of the AGGG (Advisory Group on Greenhouse Gases). This is discussed further below.
[4] Once individuals selected for participation in the IPCC have been nominated by governments, the Working Group Bureau of the WMO then selects from the list of government nominations.

and non-governmental is highly permeable and often very blurred. This does not preclude looking at what may still be considered principally non-state actors that identify themselves more strongly as part of a scientific community than as representatives of a particular government.

The impact of WG1, in common with the epistemic communities approach, can be assessed by the degree to which it has been able to nurture the 'reformulation of national objectives' and encourage states to 'realise that new attitudes and political decision-making procedures are necessary' (Haas 1990a:xxi) to address the problem of global warming. In more specific terms, influence can also be traced to the impact of views concerning the kind of gases to be controlled, the methods to be used in measurement and projection, and the rules to be employed in directing policy (Haas 1990b:350). Yet a key methodological difficulty identified by Vogler (1995:205) is that, 'Because epistemic communities are relatively amorphous and because their claim to significance is based upon the engineering of cognitive change amongst policy-makers, their real impact is difficult to establish.' In this respect the significance of the time lag between the giving of advice by WG1 and resultant policy action may be too long to be detected by a project that covers the first six years of WG1. Jaeger and O'Riordan (1996:17) note, for example, that despite the importance of the conclusions drawn in WG1's *Second Assessment Report* (IPCC 1995), 'it is still unlikely to move the political machinery, at least for a few years yet, into any significant new initiatives'.

These methodological difficulties mean that longer-term government learning is not captured in this analysis; shorter-term government reactions to advice inevitably prevail. There is also the danger of attributing policy-makers' use of ideas to influence on the part of the community. Whilst they may be introduced into the debate in a certain way by epistemic sponsors, they are subsequently transformed by the policy process, and the form in which they are accepted by policy-makers may be so distorted as to make a direct connection back to their sponsors, problematic. This undoubtedly introduces further analytical difficulties, but the fact that the work of the community remains the validating reference point during the later stages of policy development, is testimony to their ability to project the importance and legitimacy of their work in policy fora.

Sonia Boehmer-Christiansen (1996) supports her argument about the influence of the IPCC with the counterfactual argument that policy-makers' acceptance of the WG1's emphasis on research has gone further than would have been the case if governments had been left to their own devices. Haas (1990b, 1992) also develops his argument on a speculative basis. He notes, for example, that without the presence of the epistemic community it is likely that there would have been less cooperation, enforcement would have been slower and less aggressive, and variation among efforts would have been much wider (Haas 1990b:358). Consistent with both the general approach in this book and with the type of analysis employed by these writers, a combination of conceptual and empirical material and examples of observable and tacit/non-observable influence will be drawn together to understand the nature of the influence of WG1.

The following sections assess the impact of epistemic communities at different stages of the policy process on global warming.

3.3   Agenda-setting

Before choices involving cooperation can be made, circumstances must be assessed and interests defined. (Adler and Haas 1992:367)

A number of far-reaching claims have been made of the IPCC, and WG1 in particular. The emergence of an international regime to confront global climate change is accounted for by some writers by the 'increasing realisation among scientists of the potential scale of the problem and their increasing internal consensus about the definition of the problem' (Paterson 1992b:175). Boehmer-Christiansen (1995a:1) also suggests 'global warming could not have entered international politics without the support of influential voices from the scientific community'. Hempel (1993:230) notes that, 'Although science has played a major role in past environmental policy, it has never influenced environmental agenda-setting more rapidly and forcefully than in the case of atmospheric change.' The influence of WG1 in setting the terms of political debate on the issue seems to go unquestioned.

The ability of scientists to translate a problem into an issue requiring an institutional response provides them with significant leverage in the policy debate. This agenda-setting influence derives in part from the fact that the ability of states to identify solutions and interests will 'first and foremost depend on the degree of scientific certainty' (Paterson 1994:293). Complexity and uncertainty induce policy-makers to seek greater clarity and predictability through consultation with expert advisers. In this context, Hempel (1993:231) observes, 'Politicians still choose one policy option over another, but it is increasingly the experts who shape the deliberative framework within which they must choose.' This builds on Haas's (1992:2) point that 'How states identify their interests and recognise the latitude of actions deemed appropriate in specific issue-areas, are functions of the manner in which the problems are ... represented by those to whom they turn for advice under conditions of uncertainty.' In other words the epistemic community derives power from its ability to make intelligible and comprehensible (for policy-makers) a problem previously understood only by specialists. The condition of uncertainty is also significant because Haas suggests that scientific advice may have been crucial whilst the 'veil of uncertainty' existed, before 'evidence of winners and losers became apparent' (Haas 1990b:360). In other words, by reducing uncertainty WG1 is in many ways laying the groundwork for the demise of its privileged input into the debate, because policy-makers come to depend on its advice less strongly once their interests become clearer.

At the same time, however, the political impact of WG1's work is heightened by the perceived degree of consensus among its members. Shackley (undated) argues that the appearance of consensus is central to the authority and legitimacy of the IPCC as an institution. WG1 has produced a number of reports that reflect a degree of consensus about the seriousness of the threat posed by climate change and the nature of action required.[5] Lunde (1991b:48) quotes WG1 chair John Houghton as saying, 'The scientific consensus on global warming (as portrayed by IPCC in May 1990) is close to unanimous. Less than 10 leading scientists disagree with the main findings of the IPCC's scientific assessment.'[6]

Lunde (1991b, 1991c) applies a series of criteria to assess the robustness of the IPCC 'consensus', including its representativeness, political legitimacy (in terms of whether it is subject to 'unsound' [sic] external pressures) and the extent to which it reduces uncertainty. He finds that most attempts to discredit the way in which the conclusions of the IPCC have been arrived at are either unfounded, or have success-fully been deflected. This is important, given Haas's observation that a 'fracturing of [the epistemic communities'] shared causal understandings of the world through the invalidation of its causal beliefs would lead to a weakening of its authoritative position' (P. Haas 1990a:57). It can be argued that one way in which the outward pre-sentation of consensus has been maintained, is through the 'deliberate exclusion of particular scientists who held deeply hostile views to those of the majority of climate scientists' (Paterson 1994:235).[7]

Similarly, the uncertainty range in WG1's assessment was left deliberately large so that most researchers could support the consensus (Boehmer-Christiansen 1996). This is particularly apparent in the policy-makers' summaries, which Boehmer-Christiansen (1995a:3) describes as 'skilful exercises in scientific ambiguity'. The extent of the portrayed ambiguity is demonstrated by the fact that the content of the 1990 assessment report simultaneously allowed Greenpeace to call for a 60 percent reduction in CO_2 emissions and the UK Treasury to conclude that no action was necessary until more scientific certainty developed, each citing the same source (Boehmer-Christiansen 1995a:3). The pressure to construct a body of advice amena-ble to all policy users appears to have affected the nature and tone of the advice.

Sluijs et al. (1998) also show how the IPCC has sought to consistently maintain validity of a climate sensitivity range of 1.5°C to 4.5°C over two decades, even

[5] The preface of the IPCC's First Assessment Report states that it is 'the most authoritative and strongly sup-ported statement on climate change that has ever been made by the international scientific community' (IPCC 1990: preface). The report states, for example, '*We are certain* of the following: ... Emissions resulting from human activities are substantially increasing the atmospheric concentrations of the greenhouse gases. These increases *will* enhance the greenhouse effect, resulting in an average warming of the earth's surface' (IPCC 1990:xi, emphasis added).

[6] For criticism of this assertion of consensus see the Leipzig declaration (1996) of the Science and Environmental Policy Project and the European Science and Environment Forum's 'Global warming: The continuing debate' (1997).

[7] The so-called Phoenix group is an example of a group of fourteen scientists critical of the greenhouse consen-sus who were 'kept out' of the IPCC process (Shackley undated:5). Mikhail Budyko, who argued that green-house warming should be accelerated, is an example of an individual who was effectively sidelined from the IPCC process for holding what were considered to be extreme views (Lunde 1991c).

though the range of individual GCM results has changed over time. As well as acting as an anchoring device' (ibid.),[8] the apparent consensus around this estimate 'reflects an implicit social contract among the various scientists and policy specialists involved, which allows "the same" concept to accommodate tacitly different local meanings' (ibid.:291). The impression that scientific disagreement or uncertainty compromises policy authority and effectiveness, encourages the expert community to present as high a degree of certainty and consensus in their work as possible. Indeed, in delivering a statement on behalf of the IPCC to the first session of the climate negotiations, Bert Bolin used this sensitivity range to claim that the scientific community is confident that no climate change is an impossibility (Sluijs et al. 1998:309). The process is an ongoing one where experts must simultaneously negotiate support for the work with scientific peers and the policy community (ibid.). As the issue progresses through the policy cycle, so the demand by policy-makers for evidence of certainty, consistency and robustness becomes more pressing (ibid.).

In many ways, then, it is not the actual degree of scientific consensus that exists among the members of the community that is most important. It is the perception among the users of their advice (policy-makers) that the knowledge is sufficiently authoritative to form the basis of policy.[9] The message emanating from the work of WG1 is thought to be clear by both the IPCC and those outside it (Boehmer-Christiansen and Skea 1994). At the Second World Climate Conference in Geneva in 1990, the IPCC's assessment was accepted by over 137 attending countries as the scientific basis for negotiations towards a convention (Sluijs et al. 1998). This proposal was endorsed by the United Nations General Assembly in December 1990. The IPCC's report was regarded as sufficiently authoritative to prompt fourteen of the OECD's twenty-four member states to initiate policies to stabilise or reduce their levels of greenhouse gas emissions (Rowlands 1995:89).

The willingness of the international community to accept the advice of the IPCC can be contrasted with the marginalisation of the AGGG (Advisory Group on Greenhouse Gases). The predecessor of the IPCC, the AGGG, has been described as a 'large and increasingly influential epistemic community of climate change scientists and supporters in international organisations during the 1980s' (Vogler 1995:204),[10] but 'The energy policy implications of [its] early advice led to the demise of the AGGG' (Boehmer-Christiansen 1994b:185). According to Lunde (1991c:35), 'the field would probably have been left to the scientists had not the political affectedness been that strong'. Without dwelling on the history of the AGGG, the important point here is that it was the agenda-setting capacity

---

[8] Anchoring is thought to be the unintentional product of the political interplay between the peer review process and policy treatment of uncertainty in scientific knowledge (Sluijs et al. 1998).
[9] The IPCC represents a major part of the international scientific expertise on the issue (Isaksen 1993:77) and the world's leading climatologists as recognised by the governments that convened the panel in 1988.
[10] The AGGG was formed in 1985 at Villach, Austria, under the leadership of figures such as Bert Bolin and William Clark of Harvard University.

of the group and the overtly politicised nature of their advice that encouraged governments to assert greater control over the source of their advice by creating the IPCC.

This emphasises the primary role of states in setting the scope for the participation of experts in policy debates, as well as their willingness to intervene in the process of knowledge accumulation and exchange if it is thought to touch upon politically sensitive areas. As Shackley (1997) notes, the debate has become far more conflictual now that the IPCC has moved from studying the relevant phenomena to considering policy responses. At the IPCC meeting in the Maldives in September 1997, the US, China and Russia raised objections about governments' loss of control over the formulation of the synthesis report, which will ensure line-by-line vetting of the text of the Policy-Makers' Summary by government delegations. Governmental control is also exercised through the election of WG1's and vice-chairs. At the same meeting, five new vice-chairs had to be created to avoid battles between countries over who should be elected. The appointments are hotly contested, because while amendments to the main reports proposed by government delegations and NGOs are discussed at plenary meetings of WG1, lead authors have the final say on the precise contents of the reports. Though objections about abuse of IPCC procedures have been raised by 'laggards' such as Australia, government trust in the scientific legitimacy of the IPCC process continues to be strong; a factor critical to the IPCC's success (Shackley 1997).

The fate of the AGGG and successive attempts by governments to set the boundaries within which the IPCC operates can be argued to have tempered the political assertiveness of the IPCC. It is perhaps awareness on the part of WG1 of the political repercussions of its advice, that it has shied away from the task of defining what constitutes a dangerous level of human interference with the climate system. Avoiding such interference provides the very goal of the Climate Convention, yet at a special workshop on Article 2 of the convention (Fortaleza, Brazil, October 1994) the IPCC concluded that it was beyond its brief to define what might constitute dangerous climate change, since that would depend upon economic and political objectives (Parry et al. 1996). The panel's anticipation of negative government reaction to their advice discourages them from voicing their opinion on such a controversial matter.

It has also been suggested that more strongly worded warnings on climate change were subject to an internal veto, on the basis that they would probably be rejected outright by politicians. Wynne (1995) shows how down-playing the more alarming end of potential temperature increases in order to manufacture a set of scientific conclusions suitable for policy, has been an ongoing characteristic of the science–politics interface on global warming. Leggett (1990) and Lunde also identify a sense in which 'IPCC scientists have been forced by politics, or other forms of social pressure, to play down the greenhouse gravity of their findings' (Lunde 1991c:15), aware of the reluctance of governments to take decisive and potentially costly action.[11] In this regard, WG1 can be argued to be in the business of offering a 'serviceable truth' (Hannigan 1995:91).[12]

A draft of the third assessment report of WG1 is to be circulated among interested parties in order to pre-empt or avoid many of the controversies that engulfed the second report. By voicing their reactions to different formulations and interpretations of the science during plenary sessions, parties provide a useful barometer of opinion that the IPCC can internalise in order to facilitate wider acceptance of its final findings and 'avoid political resistance from higher level sources within national governments' (Shackley 1997:78). This allows scientists to 'test out' the IPCC's knowledge claims in the quasi-political laboratory of the IPCC's plenary sessions' (ibid.). In this sense the political community is implicated in the formulation and construction of the science in direct and indirect ways, so that the process is one of accommodation and negotiation rather than the linear transfer of advice between two discrete communities (ibid.). Indeed it is the policy-makers' summary which is most widely circulated and referred to in the policy debate. The fact that it bears the imprint of actors other than WG1 does not detract from the fact that the knowledge contained in the summary is their own work; and the extent of its impact is in many ways a function of WG1's willingness and ability to project its findings in policy fora. The plenary sessions can contest, and on occasion successfully reject, particular parts of assessment, but the process of encoding, framing and filtering has already taken place by this stage. The range of interpretative packages available to parties has been narrowed by the assembling process.

The impact of WG1 is therefore more subtle than the direct provision of advice. During the initial phase of defining the nature of the problem, 'certain phenomena are identified as facts and set into a given causal framework that will denote the implications of such "facts"' (P. Haas 1990a:52). Other interpretations are, by definition, excluded and kept from further discussion in a policy context. The operation of these 'filters' conditions the subsequent course of debate. Jaeger and O'Riordan, quote IPCC chair Professor Bolin, arguing that the role of the scientist is to 'delineate a range of future opportunities and analyse what the implications of development along one course or another might be [but] not to recommend one or the other' (Jaeger and O'Riordan 1996:3). Despite Bolin's claim to political neutrality, however, it is possible to argue that the selection of particular greenhouse gases as targets for policy action, for example, is politically significant, given that focus on different gases implicates the apportionment of burden and responsibility upon particular states.[13] Also, even though the climatological

---

[11] Cain (1983:97) notes in relation to the history of the WMO, for example, that 'cooperation exists and action moves forward as long as the WMO deals with technical activity and avoids addressing policy-related matters'.

[12] Again in relation to the point about internalising the expectations of the policy community, Lunde (1991c:79) notes that as early as the Villach conference 'scientists seem to have been strongly aware of the potentially important implications of the message they were about to convey to the outside world'.

[13] For example focus on carbon dioxide heavily implicates the industrialised countries of the North and disadvantages those states which are dependent upon reserves of fossil fuels, principally oil, coal and gas. The focus on methane, however, implicates less developed states, which emit this gas during agricultural production.

community has shown itself to be unwilling to identify a level of interference in the climate system that might be considered dangerous, it has suggested criteria against which this may be determined (Parry et al. 1996). This creates a framework of options, including some and excluding others. The community can also help shape notions of 'winners and losers', 'victims and perpetrators' that will fundamentally influence the course of subsequent negotiations. In more general terms, Liberatore (1994) argues that WG1 was influential in identifying the climate change problem as global, human-induced and a serious threat. It is this process of framing, imposing order on how the problem is to be understood and making a problem politically 'treatable', that enables the community of scientists to project their views on the future shape of political action.

In this translation process there is scope for 'knowledge brokers' (Litfin 1994) to shape the way in which knowledge is presented to policy-makers in accordance with their preferred agenda. Fleagle (1992:75) shows how WG1, 'reflecting European views', framed their advice in the policy-makers' summary of the *First Assessment Report* (IPCC 1990) in terms of a precautionary discourse emphasising that costs may be reduced by acting now. Singer (1992b:2), refers to this summary as essentially 'a document of governments not of scientists', on the basis that the policy-makers' summaries are unrepresentative of the main report and tend to exaggerate the degree of certainty that exists in the main body of the reports (ibid.:3).[14] The point is that the influence of the epistemic community is partly conditional on the way it chooses to construct its advice or assemble its findings. As NGOs attending the Second World Climate Conference noted, 'scientists who worked at Villach, Bellagio and on Working Group I of the IPCC have not only brought the facts to our attention, but have done so in a way which has forced governments to sit up and take notice' (*ECO*, issue 2, Geneva 1990).

An interesting addition to the agenda-setting argument, is the notion that WG1's greatest influence derives from its ability to emphasise the need for further research; a course of action from which it directly benefits. It would, for example, be plausible to argue that the scientific community has an interest in emphasising uncertainty to the point where governments acknowledge the need for greater degrees of climate research and hence apportion funding accordingly, raising both the profile of and the financial backing for the scientists' work (Boehmer-Christiansen 1996).[15]

Paterson (1994:232) notes that 'The motivation for international scientific cooperation [on climate change] seems to have been partly economic', on the basis that scientists tend to become involved in finding solutions to the problems they have been instrumental in identifying, and therefore develop a self-interest in

---

[14] Singer (1992b:1–2) argues that the 'By selectively extracting from the often conflicting statements in the Report that express existing doubts and uncertainties, the summary's firm tone leads policy-makers to believe that the existence of a climate problem has been confirmed by "scientific consensus".'

[15] Haas mentions 'political rewards' such as 'research contracts, regional activity centres, and new technology and training in its use' that scientists accrue from integration into the policy process. He continues, 'In order to receive such rewards they continued to attend negotiations' (Haas 1990a:218).

maintaining themselves in international bodies. The way that scientists in the IPCC have been able to cast a problem-defining role for themselves, and succeeded in ensuring that the Climate Convention emphasises the importance of scientific monitoring and increased research (helping to ensure the dependence of the international community upon their work), is illustrative of this process. To date, most international obligations on climate change consist of major data collection and planning efforts that are research-intensive and require significant input from scientific experts. According to Vogler (1995:206), 'the immediate beneficiaries of the FCCC will be the research communities themselves, for if it does nothing else the Convention points towards a continual and expanded effort in research and data gathering'.[16]

Wynne (1994:184) argues that this is the result of 'network interdependency' and 'mutual bootstrapping' between scientific and policy actors, which develops through institutional interaction in working parties and advisory committees. In the extreme, what develops is a 'capture' (Boehmer-Christiansen 1996:175) of the regime by the epistemic communities, who market their knowledge as a commodity. Part of the marketing strategy involves maintaining the idea that scientific uncertainty is a finite condition, that if more research money is invested, definitive answers will be found. This permits scientists simultaneously to claim a legitimate role for themselves by showing what they already know, whilst being careful to highlight the many other areas of research that require work in order properly to serve the policy community. This is the basis of Boehmer-Christiansen's (1994a:140) claim that the global research community on climate 'acted primarily as a lobby for its own research agendas'.

Once secured, however, it is argued that this funding, to some extent, ties researchers to their sponsors. Boehmer-Christiansen and Skea (1994) show how more than half of the members of WG1 think that the international organisations sponsoring their work (particularly UNEP and the WMO) influenced the policymakers' summary, and how more than half also believe governments to have exercised a similar influence. Once WG1 had established itself as the key source of scientific advice on the back of government funding, 'It could not afford to offend major governments or its sponsors' (Boehmer-Christiansen 1994b:190). NASA (National Aeronautics and Space Administration) scientist James Hansen learnt this lesson when he lost his funding from the Department of Energy (DoE) in the US for his outspoken comments on global warming. Lunde (1991c:42) quotes the journal *Science* on this incident: 'In short, indications that carbon emissions might have to be limited was not a message the DoE wanted to hear'. Hence the way in which knowledge is constructed and the policy recommendations that are brought to the fore are conditioned by an anticipated sense of what sponsors will

---

[16] Boehmer-Christiansen and Skea (1994) show that many (84 percent) of the actors involved in the policymaking process believe the IPCC to have played a key part in attracting financial resources for research programmes with which it is associated.

find acceptable. Referring to the IPCC, Jaeger and O'Riordan (1996:5) argue: 'it is frankly impossible for such a panel actually to remain aloof from the political processes that both shape its existence and respond to its propositions, if for no other reason than that the pattern of response itself will determine how the scientific effort will proceed'.

One of the key problems here is determining who is *influential* and who is being *influenced* when common interests are served by the same agenda. For Boehmer-Christiansen the research enterprise is an ally of a global regime that seeks to avoid substantive global regulations. She notes how, in relation to the UK, 'Government interest was in funding more research rather than enforcing changes in energy policy' (Boehmer-Christiansen 1994b:195). Bergesen (1989:124) describes what develops as an 'unholy alliance' between 'politicians wanting to postpone difficult decisions and scientists, or rather administrators of scientific institutions, preoccupied with fund-raising for their institutes'. One response to the dilemma of attributing influence is that experts' influence is conditional upon the acceptability of their claims to the users of their advice. Hence a limit to agenda-setting influence is self-censorship on the part of the scientific community in respect of the proposals or recommendations they make, with an eye to what the response of politicians is likely to be. At the same time governments, if they are seeking to avoid having to take action beyond investing in more scientific research, are dependent on the willingness of the scientific community to state that more research is an acceptable short-term solution.

Moreover, whether or not the presentation of knowledge in a way that meets policy-makers' needs is a product of the pursuit of research money or the inevitable consequence of the process of assembling comprehensible findings amid uncertainty, to some extent boils down to whether or not the influence derived from the emphasis on the need for more research is intended. And yet for the purposes of this analysis, it does not matter. The point is that whatever the motives, WG1 was able to make an impact and accrue influence as a result of presenting its knowledge in this way.

It is clear, moreover, that the agenda-setting influence of WG1 is coincident with other factors beyond its control. Litfin (1994) shows how exogenous factors shape the political salience of various modes of interpreting knowledge. Scientists are often approached by decision-makers in order that a crisis may be diffused. It is such crises that permit the integration of scientists into the decision-making process. Paterson (1994:218) hints at this when he notes that the political empowerment of the scientific community 'often requires the consensus to coincide with lay perceptions of the problem'. In this regard, the exceptionally hot summer of 1988 in the US, and the series of abnormal climatic events across the globe in the late 1980s, drew attention to the scientific work on climate change, in response to public fears about the possible effects of climate change. As Paterson notes, the 'public's sensitisation to environmental issues in general in the 1980s ... enabled it [the IPCC] to get global warming onto the agenda' (ibid.:241).

As a number of commentators have demonstrated (Boehmer-Christiansen 1996; Kellogg 1987), the state of greenhouse science has changed only very slowly over time, sufficient to say that a major breakthrough did not take place in 1988.[17] The scientific knowledge of global warming that governments had at their disposal in 1988 had not fundamentally changed for some time. When global warming exploded (given that it did so suddenly) in 1988 as a political issue,[18] it did so not in relation to the findings of the scientific community, which were in any case underdeveloped, but in response to widespread public concern about the issue, set against a backcloth of some of the hottest years on record and prolonged drought, events which the media were quick to suggest were directly attributable to global warming (Ungar 1992). Rowlands (1995:73) notes how the outspoken comments of NASA scientist James Hansen, combined with the hot weather, did more to influence Congress on the issue 'than any number of scientific treatises on the subject'.

As early as the first Villach workshop in November 1980, warnings had been issued that the accumulation of greenhouse gases in the atmosphere required an urgent response, yet at that time the political impact was negligible (Rowlands 1995:72). The participants at the Villach–Bellagio workshops of the mid-1980s moreover, had discussed and agreed upon policies to respond to climate change (ibid.). Shackley (undated) adds, 'There was a fairly strong scientific consensus on the likely effects of greenhouse gas emissions from the early 1980s onwards, yet the problem only surfaced onto the wider policy agenda in the wake of the American drought of 1988.' The warming trend lent credibility to the views of some in the emerging epistemic community (who went on to form WG1) that precautionary action was desirable. An opportunity was provided, therefore, for the frame of interpretation being propagated by some within the community to acquire salience. As Litfin (1995:276) notes, 'severe drought during the summer of 1988 lent credibility to a precautionary discourse on climate change'. Boehmer-Christiansen (1995a:4) concludes that 'the rapid politicisation of the climate debate occurred in a context of scientific ambivalence, influenced by forces beyond the control of science'. Similarly the 'greenhouse backlash' (Pearce 1995d) that developed against the work of the IPCC in the early to mid-1990s has to be understood against a background of media 'greenhouse fatigue' (ibid.:1995a), the low price of oil and the reassertion of government opposition to action on the issue, particularly in the US with the rise of Republican dominance in Congress.

Haas acknowledges to some extent the importance of a wider context in creating political space for the advice of the scientific community where he notes that 'If perceived uncertainty by politicians is high and public pressure is severe, then epistemic communities may be effectively able to promote policies that are further from the political "norm" ' (Haas 1990b:352). The reactive policy-making environment that can be created by exogenous shocks and subsequent public concern means that,

---

[17] In fact Kellogg (1987:132) argues that by the far the biggest advances in the evolution of awareness occurred during the 1970s. The political climate at that time was not receptive to policy change, however.

[18] The year 1988 is seen by most writers on global warming as the turning point in terms of the perception of the international community of global warming as an issue deserving of a coordinated international political response (Brenton 1994; Paterson 1992b:155; Rowlands 1995).

as Haas notes, 'In addition to enabling one to reduce uncertainty, resorting to expert advice is valuable domestically as a political device that helps one avoid or postpone short-term conflict by shifting policy responsibility to the experts' (ibid.:54). The community can be used 'to pin the blame for a policy failure or simply as a stop-gap measure to appease public clamour for action' (Haas 1992:42; 1990b:350). Boehmer-Christiansen, for example, shows how UK climate scientists' advice came to be in demand once it became clear that many of the proposed solutions to global warming would require a degree of government intervention, which the government of the time was anxious to avoid. Further scientific research provided justification for delaying interventionist policy strategies (Boehmer-Christiansen 1995b:3).

The degree of scientific certainty required of climate experts differs from state to state, according to the political costs that attend responding to the issue. The US has continually emphasised the need for a greater degree of scientific certainty before taking further action, the Netherlands less so. Questions of energy infrastructure and culture, and the political weight of corporate lobbies help to explain the former position, and vulnerability to climate impacts and an influential environmental lobby the latter. Boehmer-Christiansen (1996) and Rowlands (1995) note how WG1 was able to exert a great deal of influence over its host nation: the UK. Rowlands (1995:79) in particular notes how 'The UK government was quick to respond to the interim report's publication ... its officials changed their position on the science.' The advice of WG1 was useful in furthering a range of predefined goals in the UK context relating to bureaucratic turf battles and the advancement of a more positive international position (Boehmer-Christiansen 1995b).

In contrast, despite receiving advice from its scientists in the early 1990s to the effect that greenhouse warming posed a potential threat sufficient to merit a prompt response, the US continued to highlight uncertainties in the science as a basis for delaying action (Rowlands 1995:80). Hence if the epistemic community only manages to influence the position of one country, or of a particular group of countries, 'then its influence is merely the function of that country's or body's influence over others' (Adler and Haas 1992:379). The failure of WG1 scientists to influence the US position meant that a key player in the international debate remained outside its sphere of influence. The ability of WG1 to set the agenda varies quite considerably from state to state.

Assumptions in the epistemic community literature about bureaucratic access as a source of political influence are also problematic. As Haas notes, 'In the US atmospheric scientists who actively believe in the need for prompt carbon dioxide controls are barred access to the administration', so that 'the community's avenue to decision-making in some of the major actors is blocked' (Haas 1990b:359). The Environmental Protection Agency (EPA, the bureaucratic arm in which the epistemic community was more entrenched) was easily marginalised in the US decision-making process (Litfin 1994). It is important to note that a community may be regarded as an authority in Environment Departments, but if its claim to authority is not recognised elsewhere in the government its political influence will be restricted.

The extent of the hostility of the US Office of Management and Budget (OMB) to the tone of expert warnings about climate change was made visible in May 1989, when the office was forced to admit that it had altered the congressional testimony of climate change action advocate James Hansen (Colglazier 1991). The 'learning space' seems to differ from department to department and the ability of departments to influence the overall shape of policy also differs. Litfin's (1995) argument that the influence of knowledge brokers derives from the plausibility of the interpretations they advance (in terms of both their consensual backing and their political palatability), the loudness of their voices (in terms of overall bureaucratic influence) and the political context in which they operate, comes close to describing appropriate criteria for the influence of WG1.

The conditions attached to the ability of the epistemic community to set the agenda on global warming seem quite considerable. Nevertheless some role in defining the scope of the problem can be attributed to WG1. As Paterson (1994:231) observes, 'Their identification of the problem, their active fostering of a consensus on the nature of the problem, and their agency in pushing for a political response, were all important in explaining why global warming became an issue high on the international agenda.' There does, however, appear to be a tendency among states to 'read off' from the scientific advice they receive, only those policy implications which are considered to be politically palatable.[19] This point is developed in the next section, which discusses the influence of WG1 at the international negotiations on climate change.

3.4 Negotiation-bargaining

For Haas, international negotiations form part of a continuing process of reducing uncertainty, where leaders defer to experts over the issue in question (Haas 1990b:350). Scientists who sit on negotiating delegations are able to provide on-the-spot advice as the negotiations proceed, providing 'specific bits of advice regarding the scope of collective arrangements under consideration' (Haas 1990a:56). In precise terms, this can take the form of helping to draft documents (ibid.:225). If members of the same epistemic community belong to a range of negotiating delegations, policy convergence among the different parties may ensue. In more general terms, Haas argues that 'Meetings at which an epistemic community is well represented would be more constructive than those in which it is not' (ibid.:56).

In order to influence a range of delegations, the IPCC has to be viewed as truly transnational in terms of its representation. This helps to avoid situations where scientific results produced by institutions in one country involved in the political

---

[19] As Adler and Haas (1992:381) note, 'the decision-makers primary goal of soliciting advice from an epistemic community may be the political goal of building domestic or international coalitions in support of their policies'.

process are not recognised as valid by other countries (Bergesen 1989:124).[20] The need, for example, for North–South cooperative scientific research is particularly pronounced in the case of global warming as borne out by the existence of the Special Committee on the Participation of Developing Countries, which was created in 1989 to enhance the involvement of scientists from less developed states (Rowlands 1995).[21] There have also been attempts to appoint a developing country scientist as one of the lead authors for each chapter of the IPCC reports, as well as to provide financial support to assist their involvement. Transnational representation enables the community to be regarded as authoritative and less coloured by the perspective of any one group of countries, by a greater number of parties to the convention and therefore to extend the breadth of its influence.

WG1 cannot claim to be genuinely geographically inclusive however, because the expertise it tends to value is not evenly distributed globally (Shackley 1997:78). It remains the case that out of the twenty-eight authors of the technical summary of the Second Assessment Report, only three were from less developed countries (one from Brazil and two from Kenya) and all convening lead authors of chapters in the 1995 assessment were from OECD countries.

Bearing all this in mind, this section traces the empirical history of the interaction between the advice of the scientific community and its impact upon global climate policy at the negotiating stage of the policy process.

The input of WG1 during the negotiating stage of the policy process was marginalised by the creation of the Intergovernmental Negotiating Committee on Climate Change (INC) to manage the negotiations on climate change from February 1991 onwards. This was thought to deliver 'the policy debate fully into the hands of diplomats' (Boehmer-Christiansen 1994b:191). Part of the reason for the replacement of the AGGG by the IPCC in the first place was pressure from the US State Department, with the support of the Department of Energy (channelled through the executive committee of the WMO), to keep the scientific assessment in government hands. The move was a reaction to the political pace that had been set for the issue by the AGGG's organisation of both the Toronto conference and the Second World Climate Conference (SWCC) (Boehmer-Christiansen 1996).[22] The creation of both the INC and the IPCC can be argued, therefore, to represent attempts by government officials to take charge of the direction of policy as the political implications of action became clearer (see Section 3.5).

---

[20] Most of the process by which scientists have generated and developed their knowledge has been transnational, despite the predominance of input by Western scientists (Rowlands 1995). Boehmer-Christiansen (1994a) notes, moreover, how most of the reviewers for WG1 had personal ties with global programmes such as the WCRP (World Climate Research Programme), the IGBP (International Geo-Biosphere Programme) and UNESCO's (United Nations Educational, Scientific and Cultural Organisation) Oceanic Commission.

[21] Cain (1983) also notes how the ICSU (International Council of Scientific Unions) has played an important part in supporting multigovernmental climate-related efforts.

[22] Boehmer-Christiansen (1994b:187) notes that 'the AGGG proposals and claims had come from institutions which, to the US government, represented a lobby it deeply distrusted'.

It is also clear that the conception of what constitutes an epistemic community broadens at the negotiating stage. Vogler (1995:204) notes how like-minded politicians and officials can be considered part of the epistemic community in the negotiations.[23] This broader conception of an epistemic community, which includes government officials and scientists who share common objectives, can be witnessed in the aftermath of the Villach conference of 1985, when the Environment Ministries of Canada, Sweden, the UK, Austria and the Netherlands helped to disseminate the findings of the scientists' conference (Boehmer-Christiansen 1994b).

The function of WG1 in the negotiations was to set the goal 'from which should be derived the necessary social, economic and other policies for survival' (Wynne 1994:171). Scientists had been pre-eminent between 1985 and 1988 in laying out what they considered to be an appropriate regime for handling climate change at the international level. At the Toronto conference on the changing atmosphere in 1988 and the Villach–Bellagio workshops before that (1985 and 1987), emphasis was laid by the attending scientists upon North–South asymmetries in terms of emissions and historical responsibility, the need for a framework convention and for the convention to have as a preliminary goal the stabilisation of concentrations of greenhouse gases in the Earth's climate system. Clear emphasis was also laid upon the need for international cooperation in the management and monitoring of research on global climate change. These dimensions of the issue emerged as key pillars of the draft conventions that were circulated in the regime's earliest stages.

Yet it is difficult to assess how far the fact that debate centred on the need for a framework convention was a response to expert advice, or prompted by a desire to pursue the issue along similar lines to the ozone question, especially given the perceived success of the ozone negotiations (Benedick 1991a). Similarly it is difficult to attribute the emphasis in the FCCC upon information exchange and the need for further research directly to the recommendations made by elements of the scientific community back in 1988 (Boyle and Ardill 1989), given that these sorts of recommendation underlie many conventions on environmental issues. The emphasis upon the differentiated obligations of the countries of North and South, reflected in the statement 'industrialised countries must implement reductions even greater than those required on average, for the globe as a whole', is also adopted in the convention. However, as Paterson (1996) argues, little influence can be directly attributed to the epistemic community for this, given that differences in emissions are so apparent as to require acknowledgement, irrespective of the guidance of WG1 per se.

The alarmist tone of the scientific advice propagated at the Toronto Conference on the Changing Atmosphere in 1985 was soon lost in the emerging policy discourse on the issue. The Toronto declaration, drawn up for the large part by scientists, spoke of an 'uncontrolled globally pervasive experiment' with consequences 'second only to a global nuclear war' (Boyle and Ardill 1989:Appendix 2). The predictive

---

[23] Haas (1992b) also includes the negotiator Richard Benedick in the epistemic community that he sees as existing on the issue of ozone depletion.

tone of the statement is assured: 'Far-reaching impacts *will* be caused by global warming' (ibid., emphasis added). The conference was clear in its call for a world atmosphere fund, financed by a levy on fossil fuel consumption, for energy growth in the South to be compensated by reductions in the North, the need for a 50 percent reduction in the emission of greenhouse gases in order to achieve climate stabilisation, and for an interim target of a 20 percent reduction of the 1988 levels of CO_2 emissions by the year 2005. The vague reference to 'common but differentiated responsibility' in the convention illustrates a significant watering down of scientists' recommendations with respect to the ways in which North–South differences might be handled and the sorts of policy that might be appropriate to redress these imbalances.

Several years later, and despite the advanced understanding of climate change demonstrated in WG1's first report, leading greenhouse emitters were 'teaching the rest of the world how to orchestrate delay' at the SWCC in 1990, according to some observers (*ECO*, issue 1, Geneva 1990). The scientists' report to the SWCC went further than WG1's first report (1990) in calling for countries to take 'immediate action' to reduce the risks of climate change, arguing that 'remaining uncertainties must not be the basis for deferring societal responses to these risks' (Jaeger and Ferguson 1991) and setting in train a precautionary norm. The report is explicit in its call that 'nations should now take steps towards reducing sources and increasing sinks of greenhouse gases', and points clearly to the goal of halting the 'build-up of greenhouse gases at a level that minimises risks to society and natural ecosystems' (ibid.). Moreover twice as many scientists contributed to this report as to the first one.

To the dismay of US scientists in the IPCC, President Bush's response to the report was 'My scientists are telling me something different to that' (*ECO*, issue 7, Geneva 1990). One senior official from an OECD country commented that the president could only have been talking about one scientist; White House Chief of Staff and greenhouse sceptic, John Sununu (ibid.). Sununu was responsible for the fact that the 'sceptical' view on the state of greenhouse science gained far more salience in White House deliberations than the predominant IPCC view. Hempel (1993) describes this as a 'tyranny' of a small group of scientists who had been supporting and legitimising White House policy. Bill Hare, then of the Australian Conservation Foundation, put it the following way: 'When science provided what sounded like convenient uncertainties, Mr. Bush and his camp followers demanded more science. Now science is just being ignored' (Hare 1990).

As *ECO* commented at the time, 'before the ink was dry on the science report we witnessed the unedifying spectacle of negotiators ignoring the science and preparing to do nothing by gutting the draft ministerial declaration to remove talk of targets and dates' (*ECO*, issue 7, Geneva 1990). The draft ministerial statement makes no reference whatsoever to the scientific statement of the conference, leaving civil servants to work in a 'scientific vacuum' (*ECO*, issue 8, Geneva 1990). The fact that the scientific statement implied that greenhouse gas reductions were needed in order to stabilise concentrations ('a continuous world-wide reduction of net CO_2

emissions by 1% per year starting now would be required') (*ECO*, issue 9, Geneva 1990) was overlooked in the initial policy response, which mentioned the stabilisation of emissions and not *concentrations* of emissions (Moss 1995). Greenhouse gases can be stabilised while the build-up in concentrations continues. For many, then, the Second World Climate Conference in Geneva in November 1990 effectively marked the demise of the IPCC's earlier influence, and the handover of the issue from scientists to politicians (*ECO*, issue 1, Geneva 1990). The malleable nature of the advice of the scientific community to justify all manner of political positions on the issue became apparent. The interests and positions of states were becoming increasingly inflexible to the advice of the community.

In February 1991 negotiations began towards a Framework Convention on Climate Change. For some, the fact that the negotiations became necessary at all is attributable to the efforts of WG1 in emphasising the need for political action. Brenton argues that 'It is difficult to overstate the achievement of IPCC WG1 in *forcing* governments to focus on the climate change issue and participate seriously in the negotiations' (Brenton 1994:193, emphasis added). Likewise Jaeger and O'Riordan (1996:15) argue that 'On the basis of the strong statements made by IPCC WG1 ... negotiations for the climate convention began.'

The overall objective of the convention was to some extent determined by WG1, inasmuch as the treaty has as its aim the 'stabilisation of greenhouse gas concentrations in the atmosphere at a level that would prevent dangerous anthropogenic interference with the climate system' (UNFCCC 1992:Article 2). This is a scientifically determined goal and the extent to which action achieves this goal will, on one level at least, be scientifically determined. As Boehmer-Christiansen (1996:189) notes in support of her argument that the IPCC has acted primarily as a lobby for its own research interests, 'The very objective of the FCCC depends on further scientific evidence.' On this basic level, then, the convention reflects the causal beliefs of the community.

Consensus on the importance of stabilising concentrations of greenhouse gases in the atmosphere was 'rooted in the scientific evidence available at the time' (Moss 1995:3). The increasing emphasis, from 1990 onwards, on not awaiting greater scientific certainty before acting (Jaeger and Ferguson 1991) can be argued to have contributed to the establishment of a 'discursive norm in favour of precautionary action' (Litfin 1994:194), so that the entrenchment of the precautionary principle in the convention may be a further indication of WG1's influence. The consensus among the community in respect of this goal had been developing for some time. The Scientific and Technical Declaration made at the SWCC in 1990 stated, for example, that 'The long-term goal should be to halt the build-up of greenhouse gases at a level that minimises risks to society and natural ecosystems' (SWCC 1990:para 2; Jaeger and Ferguson 1991).

WG1 had also suggested that the convention should 'recognise climate change as a common concern of mankind and, at a minimum, contain general principles and obligations' (IPCC 1990:5). This too seems to have had a bearing on the convention in that the preamble emphasises the threat to humanity posed by climate change

(UNFCCC 1992:preamble). Although it is difficult to tell how many of these provisions would have been included in the convention in the absence of WG1's interventions, it is clear from Boehmer-Christiansen and Skea's (1994) work that WG1 is believed to have had a large impact on the convention, particularly among government representatives.[24] Brenton (1994:194) suggests that in the absence of WG1's forceful first report, it would be 'very difficult to see any substantive convention having emerged at all'. Pearce (1995b) similarly asserts without reservation that 'Scientists made the Climate Convention.'

These general observations aside, it is possible to argue that it was issues of political expediency that forced the pace of the negotiations towards a convention and help to account for its content. Examples include the high public expectation that a convention would be ready for signature at United Nations Conference on Environment and Development (UNCED), US election year politics, the resignation of White House Chief of Staff John Sununu, the accommodating UK position on binding targets, and a whole series of other political and economic factors that converged to quicken the pace and create intense pressure for an agreement. Indeed the process leading up to the convention's creation and final adoption was largely not driven by the advice of WG1. As Paterson (1993a:178) notes in relation to the ongoing negotiation process, 'scientific developments ... are now not particularly important because the importance of scientists was dependent [on the] relatively low level of politicisation of the issue' in the early stages of the regime. Hence despite the insertion of wording in the convention emphasising concerns voiced by WG1, political trade-offs and institutional factors seem to offer more by way of explanation at this stage of the policy process.

There also appears to be a great divergence between the recommendations of WG1 on the one hand, and what was actually included in the final FCCC (or has subsequently been discussed by way of obligations) on the other. Litfin (1994:193) notes that 'Despite the apparent existence of a powerful epistemic community of scientists, environmentalists and political leaders in favour of regulatory measures, such measures have yet to be adopted.' The IPCC has repeatedly stated that more radical measures are needed than have been proposed to date (Greene and Salt 1994:2). IPCC Chairman Bert Bolin reiterated at INC9 in February 1994 that the commitments in the convention did not go far enough (Rowlands 1995). In a Greenpeace International survey of the IPCC (all working groups), 62 percent of climate scientists expressed their dissatisfaction with progress in the climate negotiations.[25]

---

[24] More than half of the government representatives in the survey stated that the impact of the IPCC had been high (56 percent). Interestingly, the number of IPCC report writers who felt their influence to be great was considerably lower (38 percent) (Boehmer-Christiansen and Skea 1994).

[25] Eleven percent of IPCC respondents said the process was 'far too slow' and 51 percent said it was 'too slow'. Similarly 42 percent of respondents believed the work of the IPCC had not been taken seriously enough (Greenpeace International 1991b).

Moreover the Toronto declaration of 1988 had called for a 20 percent reduction in the 1988 levels of the CO_2 emissions of industrialised countries by 2005. The SWCC Scientific and Technical Declaration in 1990 made clear that such reductions were possible, and in 1990 the IPCC went on to 'calculate with confidence' that CO_2 reductions in the region of more than 60 percent would be required to stabilise the current concentrations of greenhouse gases. The caveat-peppered FCCC – free of binding obligations, reduction targets and timetables – that resulted, indicates the limits of the direct influence of WG1 upon climate policy. Even the Kyoto Protocol, which requires binding emission reductions from a number of governments and aims to achieve an overall cut in all greenhouse gases of 5.2 percent over the period 2008–12, is a far cry from what WG1 deems to be necessary.

According to the epistemic community hypothesis, the entrenchment of scientists in Environment Ministries enables them to influence the course of international negotiations. Delegations that included scientists during the global warming negotiations were mainly from developing states (Paterson 1994:237). Haas's expectation that there will be a linear relationship between a country's willingness to cooperate with the regime and the active participation of an epistemic community within its administration can be challenged with reference to the political dynamics of global warming.[26] Despite the greater representation of scientists in developing country delegations than in developed country delegations, for example, it is notable that the equity and economic issues central to the position of many Southern states had been articulated before the scientific consensus on global warming had been firmed up by the *First Assessment Report* (IPCC 1990). As Paterson (1994:240) argues, 'At best, the epistemic community provided the South with an extra intellectual basis on which to argue its case, but since the disparities in emissions ... are so obvious, no great importance can be attributed to the epistemic community on this.' Moreover, industrialised country delegations began to be dominated by Foreign Ministries, and more immediate state interests were subsequently placed above the recommendations of the scientific community. Paterson (1994:237) partly attributes this to the lack of political entrenchment of the epistemic community and therefore the ease with which it could be displaced. This shift in bureaucratic, interdepartmental power relations meant that the more influential Departments of Trade and Commerce were increasingly able to use their political weight to delay action on the issue, and to override epistemic voices.

Most of the world's leading climatologists are based in the industrialised countries of the northern hemisphere. They make up by far the largest part of the epistemic

---

[26] Paterson (1994) notes that at the 5th Session of the INC in New York in February 1992 approximately 45 percent of industrialised country delegations were from Foreign Offices and only 34 percent were from Environment or Meteorology Departments, while 22 percent of the heads of delegations from developing countries were from Foreign Offices and 47 percent were from Environment or Meteorology Departments. Although there are differences, it is not clear that this disparity is sufficient to explain the vast divergences in the negotiating stances adopted.

community represented in WG1,[27] and yet have been resolutely unable to nurture change in the policy positions of their respective governments, or encourage the realisation of new interests. As Litfin (1994:192) notes, 'scientific proficiency does not correlate with political leadership' on global warming. Governments with strong atmospheric science and climatological research capacities include the US, Canada and Australia (Boehmer-Christiansen 1996), all of which are broadly considered to be 'laggards' in the negotiations.[28]

A further factor affecting the influence of WG1 at the negotiation-bargaining stage was the presence of what might be termed policy entrepreneurs, or 'science communicators' (Susskind 1994), who enhanced the impact of the scientific conclusions of the IPCC and helped to deliver its conclusions with force. Professor Bert Bolin (chair of the IPCC), Sir John Houghton[29] (chair of WG1), Bob Watson (chief scientist of the US IPCC delegation) and Roger Revelle are examples of such individuals who significantly contributed to the emergence of scientific consensus on the issue, and played a large part in forcefully projecting the IPCC's advice. It is felt, for example, that US Vice-President Al Gore's call for emissions reductions was strongly influenced by Roger Revelle (Easterbrook 1992:24). Similarly Boehmer-Christiansen (1994b) states that the British diplomat and climate scientist Sir Crispin Tickell was a key individual in influencing the position of UK Prime Minister Thatcher on the climate change issue. David Fiske, chief scientist at the Department of the Environment in the UK is also credited with some influence upon UK climate policy, as head of delegation. However the influence that one individual can have upon the overall direction of policy should not be overestimated, and the examples (above) illustrate the limits of having a representative from an epistemic community on a negotiating delegation that is not interested in pursuing further action.

In many ways it would appear that the impact of the scientific community upon the international negotiations is, to some extent, conditional on 'the perception of the possibility of joint gains by the parties in question' (Andresen 1989:49). Initially (at the agenda-setting stage) negotiating positions are not refined and joint gains are possible. As interests become more apparent, however, that perception wanes as a sense of what is politically realistic conditions the receptiveness of policy-makers to

---

[27] Of the thirty-four lead authors involved in WG1's *First Assessment Report*, twenty-one came from the US and UK. Of the 300 contributing authors, just over half came from the US and UK. Similarly, groups capable of running GCMs are based in centres such as the UK Met Office, the National Center for Atmospheric Research (NCAR) in Colorado and the Max-Planck Institute in Hamburg (Boehmer-Christiansen and Skea 1994).

[28] The US in particular is described as being 'the best informed on the status of climate science' (Boehmer-Christiansen 1996:177). The US played a large part in initiating the IPCC process and a large proportion of US participants came from state bodies such as the EPA and NASA.

[29] Sir John Houghton is considered to be particularly responsible for the widespread perception that the WG1 had advocated immediate reductions in CO_2 emissions of over 60 percent in order to stabilise concentrations of greenhouse gases, and more generally for emphasising the degree of certainty that exists in relation to the science (Boehmer-Christiansen 1994b).

advice. It is possible to argue, moreover, that the receptiveness of governments to scientific advice is in the process of formation before scientific advice is sought, and certainly therefore, before negotiations commence. This point is illustrated by an official White House 'Talking Points' brief, distributed to US diplomats, but obtained by NGOs at the Bergen Conference in 1990. The brief included a section entitled 'Debates to avoid', which urged delegates to advance the argument that it would not be 'beneficial to discuss whether there is or is not warming, or how much or how little warming. . . . A better approach is to raise the many uncertainties that need to be better understood on this issue' (reproduced in *ECO*, issue 7, Geneva 1990).

The experience of climate policy in the EU also highlights this process. From 1986 EC energy policy emphasised energy efficiency and other 'no-regrets' measures, which went on to provide the interpretive framework for the advice of climate scientists (Liberatore 1994). In these cases, scientific advice is used to support and legitimise policies and goals that are already in place; to reinforce what is 'politically and economically palatable' (ibid.:198). These frames of interpretation are of course continually contested by new inputs of advice, but it is reasonable to argue that government positions are not as open to definition by experts as epistemic community accounts imply. What latitude does exist at this stage of the policy process may not be subject to the persuasions of scientists from WG1. The influence of WG1 with particular governments is at least partly a factor of the degree of 'negotiating space' available to states. As *ECO* stated bluntly, 'All the computer models in the world will not make a Swiss Franc of difference to governments who simply want to sell all the oil the world can be persuaded to buy' (*ECO*, issue 7, Geneva 1990). Conversely, states that benefit from the findings of WG1 are keen to project its findings in policy debates. Amelia Dulce Supertran of the Environment Ministry in the Philippines, argues that 'The work of the IPCC has a great deal of influence upon Philippino climate policy . . . we are pushing very strongly for the use of the IPCC reports because we are experiencing some of the impacts [of climate change] now.'[30]

'Laggard' states (Porter and Brown 1991) in the climate change debate have drawn upon the opinions of dissenting scientists outside WG1. In other words, the work of WG1 has not formed the key source of advice for many states. The Bush administration in particular was singled out for criticism in this respect (Hempel 1993; Rowlands 1995). As Hempel (1993:214) notes in relation to US climate politics, 'a small but influential group of scientists, all greenhouse sceptics, repeatedly advised policy-makers to deter action on the climate issue'. The availability of external challenges to WG1's interpretation of climate science, and the ability of scientists who do not subscribe to the WG1 consensus to highlight gaps and uncertainties, has made it easier for governments to cite reports that bolster policy positions that go against the recommendations of Working Group 1.

A further limitation on the influence of WG1 at this stage of the policy process is that its advice fails to map out definite ways forward in terms of policy direction,

---

[30] Interview with Supertran, Geneva, 16 July 1996.

apart from vague recommendations about sensitivity to North–South asymmetries and the need for a global instrument to tackle climate change. As noted above, WG1 has not been as proactive as the AGGG or other scientific groups were in the mid-1980s in calling for specific targets or courses of action. Indeed, as argued above, some of its influence derives from its ability to offer advice that appeals to the broadest possible constituency, achieved by making only very general recommendations. Whilst WG1 may be able to advise on the science of climate change, it is not able to inform decisions about how much global warming is socially acceptable. To return to the question of what constitutes a 'dangerous' level of climate change, Moss (1995:3) observes that 'scientists can assist in helping to identify exposure-effect relationships. ... But determination of "dangerous" is not solely a scientific process: it involves judgements about what attributes of ecosystems and human activities are most highly valued and what level of change can be considered critical.' These more overtly political questions, which the negotiations are designed to address, are beyond the scope of WG1's mandate.

This section has provided a snapshot of the influence of WG1. The advice of WG1 is continually (re)setting the agenda, as borne out by the impact of its *Second Assessment Report* (IPCC 1995). The added pressure generated by the report, which was understood to imply that the existing commitments were inadequate, may have prompted the adoption of the Geneva Declaration at COP2 (the Second Conference of the Parties to the Convention), which called for further measures in the form of a protocol 'or other legal instrument' (Newell and Paterson 1996). It may also have provided an extra impetus to the negotiations towards the Kyoto Protocol. As argued earlier, demands for particular types of advice change over time with the shifting perceptions and needs of policy-makers.

3.5 Implementation

WG1 may also be politically influential during the enforcement and verification stage of the policy process. The transnational character of WG1 disposes it well to play a significant part in pursuing compliance on behalf of international organisations. Reporting back to the IPCC and the SBI (Subsidiary Body on Implementation) from a range of host states, members of the community can help to piece together a global impression of what parties to the convention are doing and how they are going about it, set against their own scientific criteria. WG1 can help to shape the terms of review and assessment of parties' commitments by spelling out monitoring procedures that require regular review (Susskind 1994). This enables the community both to establish the benchmarks of effectiveness and to secure for itself an ongoing role in the monitoring and review process.

Uncertainty about the scientific effects of particular actions also prevails at the implementation stage, where scope for the definition of appropriate policy responses remains. Haas argues that this uncertainty means that decision-makers are ignorant about the precise effects of actions, because the scientific knowledge base may not be sufficient to offer confident predictions (Haas 1990:246). Nevertheless compliance is to some extent dependent upon information from experts, who can 'provide advice about the *likely* results of various courses of action' (Haas 1992:15; emphasis added).

Once entrenched in national bureaucracies, scientists are in a position to hasten the verification and enforcement procedure by pressuring governments to act upon their internationally agreed obligations. Haas (1989:380) finds that 'Compliance ... has been strongest in countries in which the experts were able to consolidate their power most firmly.' In relation to climate change, Boehmer-Christiansen (1995b:10) argues, 'How governments implement the comprehensive, if very imprecise strategies laid down in the Climate Convention, will remain national decisions subject to international expert advice' where scientists are consulted in the drawing up of national reports on the extent of domestic measures being taken to achieve the convention's goals. And whilst experts cannot of course dictate one direction over another, by ascribing ecological effectiveness to particular policy options they can refine the menu of response strategies.

Building on arguments developed earlier in the chapter, it is possible to explain how national policies on the fulfilment of the convention's aims reflect the interests of WG1 in terms of their emphasis on further research. The implementation of current obligations has been interpreted to imply the provision of further scientific research, which benefits both the IPCC and those governments that are unwilling to take other forms of policy action, to the extent that it would be hard to make a case for the influence of IPCC scientists predominating over actions that governments would have considered anyway. The compatibility between the recommendations of WG1 members and the policy objectives of the British government (discussed above), for example, illustrates the way in which expert advice can serve to validate courses of (in)action that governments are already set upon.

Hence it is possible to argue that the advice of the community is sought in order to help governments fulfil wider policy objectives. For whilst on the surface science is the benchmark against which to gauge the effectiveness of states' efforts towards achieving the convention's objectives, in practice the role of WG1 may be to ensure that unnecessary political sacrifices are not made. Scientific input also improves detection and monitoring mechanisms, which may help to lower the rate of defection by providing a disincentive to free-ride. Towards this end, scientists can embarrass governments by exposing poor compliance and commenting critically on the extent to which obligations have been enacted, though this is increasingly less a function of WG1 and more the role of the SBI. The effect this may have on states' self-perception of prestige may nevertheless be sufficient to encourage them to ratify the regime requirements as swiftly as possible. Although it is difficult to establish, knowledge of the ability of the community to perform this function may be internalised by states in a way that confers influence on the experts.

The duration of policies at the implementation stage is said to relate to WG1's ability to consolidate and retain its bureaucratic power (Haas 1989:380). It is difficult at this stage to comment on the extent to which the durability of policies can be attributed to the entrenchment of WG1 members in government bureaucracies, given that the first wave of commitments resulting from the convention are still in the process of being implemented. However, many of the criticisms made above about the overall influence of a community that fosters its closest ties with departments that have little influence over the overall content of policy, would also apply here.

Perhaps, especially as the costs of national action to meet international commitments become clearer, departments that have not been at the forefront of the climate debate will increasingly find that policies touch upon their bureaucratic 'turf' (such as Trade, Industry and Economics Ministries), and so the perspectives of a different set of interests, over which the members of WG1 are less likely to hold influence, become more salient. In seeking to explain the content, depth and envisaged durability of particular policy options, many other factors, which potentially have more explanatory value than the entrenchment (or otherwise) of the community in a bureaucracy, enter the equation. The nature of national responses is as likely to be influenced by economic or environmental interest groups as by members of WG1 working in government Environment Departments.

3.6 Conclusions

This chapter has provided an insight into the influence that Working Group 1 of the IPCC has been able to have over the course of the policy process to date. This influence is dependent upon politically changeable circumstances, and hence there clearly exists potential for enhanced or reduced acceptance of the scientific advice of WG1 as political coalitions and the perceptions and needs of policy-makers change.

WG1 helped to condition the negotiating frameworks that states adopted, through their initial interpretation of the problem and immediate prognosis of what constituted an appropriate policy response under conditions of uncertainty. Uncertainty was key to WG1's influence; leaders consulted experts in policy areas when it was not clear where their immediate interests lay. However, whilst scientific uncertainty undoubtedly pervaded the earliest stages of the political process, political uncertainty may not have been so present. Regardless of the timing and impact of climate change, or the exact global warming potential of particular greenhouse gases, the issue of global warming is fundamentally about the rate and type of energy production and consumption. This means that states bring to the scientific debate on climate change a relatively clear sense of which interests and activities will be affected by action, such that even initial scientific inputs will be processed by policy-makers according to their political acceptability.

Hence the assumption that there exist 'wide degrees of latitude for state action' (Haas 1992:2) does not resonate in the context of global warming, where politically palatable policy options are few. An issue that touches on the very industrial processes that sustain economic activity in industrialised countries (as global warming does) is far less likely to be acted upon, because the political stakes are that much higher. Haas (1990b:360) demonstrates some sympathy with this view when he acknowledges that, in the face of the high costs of global warming, 'technical advice and atmospheric scientists' claims of authority are surely viewed with much greater scepticism by policy-makers'.

The nature of scientific enquiry into climate change further affects the ability of experts to influence policy. The immaturity of the field of enquiry, the diversity of disciplines involved, the lack of direct imprints of climate change, added to the scale of the problem, the intrinsic uncertainties associated with climate change and the unprecedented political pressures upon the scientists,[31] all conspire to frustrate the influence of scientific actors (Lunde 1991c). Hence it is the combination of the political problem structure and the 'scientific malignancy' (ibid.:141) of the issue that make it difficult for experts to influence policy-makers.

Moreover, scientific advice is just one of many types of expert advice drawn upon by policy-makers to inform their choices. Noting the evolution of dependence on particular forms of advice, Hannigan (1995:86) argues that, 'at the policy formulation stage, the contribution of natural scientists usually diminishes while the role of socio-economic and technical experts grows'. Other types of potentially more influential epistemic advice become active in the policy debate, filtering what are considered to be plausible policy options (Rowlands 1995). For example Liberatore (1994) shows that while the scientific community played an important part in shaping the early policy debate on global warming within the EU, latterly the input of economists, policy analysts and energy technical experts has been more influential.

Underdal (1989) lists what he considers to be the conditions affecting the impact of scientific advice, including: the existence of a consensual hypothesis, the availability of a simple solution, the immediacy of the likely effects of the problem, the extent to which the issue strikes at the social centre of society,[32] the extent to which the problem develops rapidly and surprisingly, the visibility of the effects to the public, the degree of political conflict surrounding the issue, the extent to which issue linkages are prevalent and the degree to which decision-making is iterative (ibid.:259).

E. Haas (1990:42) argues that the influence of an epistemic community depends upon (1) their claims being more persuasive to dominant decision-makers than some other claim (by other experts or interest groups), and (2) a successful alliance being made with a dominant political coalition. In the former case, as has already been noted, advice on the economic effects of action on climate change and dissenting views on the science of climate change have been key influences on the position of important players. In the latter case, dominant political coalitions have been more open to advice that centres on the need for research than to advice on other forms of action that may be required. As for Underdals's (1989) criteria, the above account illustrates that political conflict is high, issue-linkages numerous in a strategically

---

[31] Andresen et al. (1994:11) argue, for instance, that 'The more intense the political controversy, the more certain the scientific conclusions have to be in order to serve as decision premises.'

[32] E. Haas (1990:171) notes that solutions offered by experts that imply obvious redistributive or regulatory measures that make visible short-term losses to important constituencies (as certain responses to global warming clearly do) are much harder for governments to accept.

negative sense (in terms of the way in which global warming has become embroiled in the North–South discourse for example) and evidence is lacking of effects that can be directly linked to global warming. These aspects of the climate *problématique* have served to lessen the impact of WG1 upon the climate policy process.

To conclude, it seems appropriate to draw on Blau's notion of 'elasticity of demand for advice' (quoted in E. Haas 1990:58). Haas uses the notion to argue that 'Governments have different elasticities of demand for scientific advice' (ibid.). This is broadly compatible with the notion developed earlier in the chapter of 'negotiating space', where, as Vogler (1995:206) notes, 'influence will vary with the extent to which consensual scientific knowledge is sought by other powerful players and the extent to which that knowledge serves to advance and legitimate other interests'. Put differently, influence will be strongest where there already exists an impetus towards political consensus (Litfin 1994) and space is provided for expert interpretations to shape as well as legitimate the agenda. This is the case both at the negotiating stage and when the policy agenda is evolving. It has been argued, for instance, that 'the uptake of climate-change science by policy-makers occurred only when the institutional and political circumstances facilitated it, namely in the mid-to-late 1980s, when a "window of opportunity" opened up' (Sluijs et al. 1998:294).

This is not to suggest that the influence of WG1 can be reduced to government's selection of those aspects of scientific advice that enable them to pursue their interests. The relationship between the expert and the policy community is more subtle than that. The influence that WG1 exerts derives from multidimensional and often symbiotic actual and perceptional exchanges with policy-makers that cannot be reduced to a one-way linear transfer of information that is either accepted or rejected.

The interaction between WG1 and policy-makers over attempts to confront global warming, suggests an agenda-setting function for the scientific community and a degree of political empowerment early on in the issue's history in terms of articulating appropriate policy goals and instruments. This took the form of bounding discussions and generating norms. The compromises that were then brokered within this broadly conceived settlement were largely uninformed by scientific expert advice, however. Domestic and other political considerations restricted the ability of negotiators to respond in accordance with the scientific advice they receive. At the implementation stage, WG1 seems to have been able to contribute to the formulation of policy both within the boundaries set by national policy-makers and by defining an important role for itself in overseeing policy implementation. The problem-structural approach would seem to suggest in general, however, that global warming is an issue too close to the 'political whirl' (Haas 1992:5) for WG1 to have a substantial and sustained influence over policy content and direction.

4

Climate of opinion: the agenda-setting role of the mass media

4.1 Introduction: agenda-setting

To be successful, a regime that seeks to forestall the onset of damaging climate change will need to address current processes and patterns of energy production and consumption, given that these are intrinsically bound up in the manufacture and release of greenhouse gases.[1] At the moment these concerns do not feature highly on the international agenda in relation to climate change. Instead of being regarded as a symptom of the unsustainability of contemporary modes of industrial production in Western states, global warming has come to be viewed as an environmental problem much like any other, whose resolution can easily be accommodated within the context of existing political and economic practices.

This chapter sets out to chart and account for global warming's presentation before policy-makers and the public alike, as a problem deserving only of incremental policy responses. It does this by discussing the role of the media as agenda-setting players in the politics of global warming. A link is suggested between conceptions of the global warming *problématique* as it is constructed by the mass media, and the nature of policy responses at the international level. In order to do this it draws on agenda-setting[2] work in the media studies literature, which explores the consequences of the mass media determining which issues, from a range of possibilities, are presented to the public. This will help to show how media coverage of global warming and the policy responses that are considered necessary are not independent and unrelated.

By looking at the importance of public opinion to international environmental regimes and then at the interaction between the media and public opinion, two models of agenda-setting will be used to describe the political impact of media coverage on climate change. The two models of agenda-setting impact are (1) direct and (2) indirect. Direct agenda-setting relates to the media's role in politicising an

---

[1] CO_2 reductions in the region of 60 percent, which are deemed by many to be necessary in order fully to address the threat of climate change (IPCC 1990; Leggett 1990), will require significant restructuring of the provision and consumption of energy.
[2] Agenda-setting is the metaphoric description of the role that the news media play in the formation of public opinion. At the heart of the agenda-setting idea is the assertion that the content of the media agenda influences public agendas in important ways (see Benton and Frazier 1991; McCombs and Shaw 1991; McCombs and Protess 1991).

issue, bringing it to public attention and generating institutional responses. Impacts can be observed on a shorter-term basis. Indirect agenda-setting refers to the framing of the debate – problem exclusion, value reinforcement and legitimation of conventional understandings – and operates over a longer time-scale. The latter trend may be harder to observe, but just as influential as more direct agenda-setting. Direct successes of the media in pushing the agenda forward are to some extent empirically demonstrable by following chains of events and recording the perceptions of decision-makers. What is harder to explore is the indirect political effect of a media-constructed frame.[3] This requires identifying the key characteristics of that frame, exploring how these relate to features of public understanding of climate change, and in turn, how such understandings interact with formal policy measures.

This chapter assesses how these functions are performed by the media in its coverage of global warming. Exploring the political role of the media helps to ascertain how particular interpretations of environmental issues acquire salience in policy; how the media exert an influence in the policy debate on global warming by projecting a particular account of the issue. It shows that there is not an *a priori* 'natural' or objective meaning that can be ascribed to global warming. Rather, it is understood as a social problem through social channels of communication that construct and make sense of it in a particular way. In regime terms, issues of problem definition are opened up and understandings of climate change can be viewed within a contested discursive terrain.

4.2 A note on methodology

It should firstly be noted that many of the studies and sources drawn upon in this chapter are based on analysis of news coverage in the West, especially the US and UK. This is partly a question of access and partly due to the bias within the media studies discipline towards these countries. Despite the limitations imposed by such a narrow focus, these states, as part of the OECD, are expected to assume a greater responsibility than many in the development of a climate change abatement strategy, given their current and historical contribution to the problem. This makes them key players in the international climate regime, and so the influences that have shaped and will continue to shape the nature of their policy responses take on international significance. Such a focus is not meant to insinuate that coverage in these countries is characteristic of coverage elsewhere. In particular the work of Chapman et al. (1997) on India shows that the issue has received different coverage in different cultural contexts and is filtered through frames that bear the imprint of distinct social and cultural circumstances.

The second potential danger in addressing a question of this nature is that an over-deterministic impression of media effects is created. In other words, any attempt to

---

[3] A frame is a central organising idea for news that supplies a context and suggests what the issue is through the use of selection, emphasis, exclusion and elaboration. (Tankard et al., quoted in Dunwoody and Griffin (1993)).

suggest a connection between popular presentations of an issue and the manner in which the issue is addressed at the international level is bound to run into problems of generalisation about public–government relations, exaggerated claims being made on behalf of the media, and a lack of sensitivity to the way different media operate and present issues. These limitations, which necessarily run through a project such as this, can only be dealt with by constantly emphasising the significance of a wider context, by adapting the scope of any conclusions that can usefully be drawn, and by being alert to the presence of a diversity of media agendas. Some media are clearly more agenda-setting than others, either because of the size of their potential audience (such as TV) or because of their access to the 'political class' (Chapman et al. 1997:43) that makes decisions on climate policy. It is this class which is most involved in the circuit of agenda formation that links media, the public and national agendas.

Associated with this is the crucial difference between causation and correlation, where correlations can be mistaken for causation. What is attempted here is not an abstract search for effects that can indisputably be attributed to media coverage of the global warming issue, however. Rather the theme explored throughout this chapter is how the interaction between public opinion and the media relates to government-level policy responses to the issue. In this respect it does not make sense to ask the question 'what effect do the media have upon climate politics?' The more interesting and researchable question is how informal constructions of the issue, apparent in mass media coverage, interact with the formal political process.

4.3   Public opinion and international environmental regimes

The effectiveness of responses to the issue of global warming will ultimately depend more on the attitudes and values of the public than on expert knowledge. The importance of public attitudes towards environmental issues is well established by writers on international environmental politics. Brenton (1994:xv) attributes the salience of the environment on Western political agendas from the 1970s onwards to the 'great upsurge' of Western popular concern about the environment. General levels of awareness are thought to explain the 'push' or 'drag' (Porter and Brown 1991) status of states in negotiating fora in respect of a range of environmental issues. According to Porter and Brown (ibid.:20), 'Public opinion . . . has had a substantial, if not decisive influence on the outcomes of global bargaining on whaling, Antarctic minerals and ozone depletion and could be a key factor in negotiations on global climate change.' Examples abound of public opinion driving the policy process.

The turnaround of Germany's position in the acid rain negotiations is partly explained by public outrage at the destruction of the Black Forest and the need for the ruling coalition to respond to public concern about ecological issues, expressed through support for the Green Party (Boehmer-Christiansen and Skea 1991:189–91). Similarly, public rejection of the use of CFCs (chlorofluorocarbons) in aerosols, if not forced change, then created negotiating space for states to agree upon emission reductions by breaking down the resistance of chemical industries, which were given an incentive to push for the internationalisation of regulations. The subsequent agreements to phase out ozone-depleting chemicals demonstrated for Brenton (1994:xvi), the importance of 'public alarm as aroused by the ozone hole'. Andrew Hurrell (1992a:402) argues that part of the reason why the issue of deforestation achieved international prominence was because 'it lent itself to dramatic and extremely effective media presentation'. He concludes, 'Amazonian deforestation then provides a particularly powerful example of the role of the media in setting the foreign policy agendas of states and in helping to shape political responses to environmental issues' (ibid.:403).

Conversely, the absence of popular concern about an issue or set of issues means that states are less likely to act on their own initiative,[4] especially in the context of an economically and politically problematic environmental issue such as global warming. Perhaps more than other environmental issues, global warming requires a high level of popular concern to stimulate a response, given the weight and influence of interests organised against further action to combat climate change (see Chapter 5). In relation to global warming, Mintzer (1992a:262) argues that public awareness will be a 'crucial vehicle for insuring the integrity and effectiveness of these international negotiations'.[5]

Clearly, numerous factors other than public opinion explain, often to a greater degree, action at the international level. The examples above nevertheless demonstrate the relevance of public opinion to international environmental policy, and suggest the importance of public opinion in the politics of climate change. The next section looks at the origin of public demands for action and sources of pressure upon governments, building on accounts that take public opinion as given without exploring the influences that help to shape it.

4.4   The importance of the media in shaping public opinion

The ways in which the public comprehend and negotiate the complexities that characterise the global warming issue, depend in important ways upon the interpretation and simplification of the issue by the media, given that much of what is understood about environmental issues is gleaned from their articulation in the mass media (Anderson 1991; Bell 1994; Bramble and Porter 1992; Underdal 1989). Understanding how the public perceives the problem of global warming, and hence the sorts of expectation that develop, requires an appreciation of the media's interaction

[4]   Caldwell (1990:17) argues that 'Initiatives for international cooperation on environmental issues have almost invariably arisen outside governmental bureaucracies' and McCormick (1989:x) also notes 'governments and policy-makers have taken very few initiatives ... instead mainly reacting to public demands'.

[5]   The extent to which other actors such as scientific communities are able to project their interpretations of a problem are also understood to be dependent, in important ways, upon the level of popular concern about an issue (Paterson 1993a). See also Chapter 3.

with public perceptions in terms of the value structures and messages it implants in popular consciousness and the more explicit 'information' it provides the public with to make sense of the problem. The media translate techno-scientific dialogues into more accessible forms of knowledge and bring international issues to the domestic realm, therefore constituting a pivotal discourse between the 'distant public' and 'local public' domains (Burgess et al. 1991).[6] How the media choose to present the significance of an issue confers meaning, context and scale upon it. In so doing, the media provide the vocabulary and tools for engagement in political discourse on global warming. In particular the uncertainty that surrounds the issue allows the media to represent and order the risk perceptions that underlie public concern.

Most writers on environmental politics, acknowledge the importance of public opinion without developing ideas about the formation and conditioning of that opinion. There are of course exceptions, and some writers have recognised the importance of the media as a communicator of environmental debates in the public realm, even if they have not pursued those themes (Bramble and Porter 1992; Brenton 1994; Kamieniecki 1993; Litfin 1994; O'Riordan and Jaeger 1996; Susskind and Ozama 1992, 1994; Weizsäcker 1994). Hurrell and Kingsbury (1992:10) note, for example, that 'The diffusion of green thinking through the workings of the global media . . . is an additional and insufficiently studied aspect of environmental politics.' Where there is mention in the literature on international politics, it is brief and connections with political outcomes are not emphasised.

Other writers, most prominently media studies scholars, have been more direct in attributing to the media the social learning that often precedes political action. It is useful here to separate direct from indirect effects of coverage. The former refers to apparent evidence of a link between widespread coverage and a change of political course, and the latter to the *frame* of coverage, which is reflected in popular interpretations of global climate change. The former might be considered a more proactive form of agenda-setting, and the latter a more passive form of agenda-setting. The following sections develop these ideas in relation to media coverage of global warming.

Direct effects

Direct agenda-setting refers to the process whereby the projection of an issue into the public arena by the media has the effect of provoking formal government activity, and connections between the former and the latter appear tenable. There is a wealth of evidence to substantiate the claim that media attention to an issue influences its perceived importance among the public (Bell 1994; McCombs and Shaw 1991). There are instances, moreover, of the media performing an identifiable agenda-setting function in relation to environmental issues.

---

[6] Burgess et al. (1991:500) define the local-public domain as 'the context of domestic lives' and distinguish this from the 'distant public' domain of 'national and international contexts of economic and political life'.

The two cresting waves of environmental awareness, the first in the 1960s and the second in the mid to late 1980s, correlate strongly with high points of media coverage of environmental issues (Beuermann and Jaeger 1996). The media are thought to have been very influential in both accelerating and directing shifts of popular sentiment on the environment (Corner 1991). It was public perception of an alarming reality that was already there – illustrated in the 1980s by, for instance, the Bhopal tragedy, the Chernobyl disaster and the continual rise in global temperatures throughout the decade (attributed to global warming) and given graphic exposure by the media – that is thought to have been the main shaper of green consciousness at a popular level (ibid.). This diffusion of concern preceded many of the more outspoken formal and documented expressions of concern by the scientific community, with Working Group 1 of the IPCC only submitting its first report on the subject in 1990. The IPCC, it is notable, was created at the height of public concern (and media coverage) about the climate change issue in 1988, perhaps indicating a political gesture in response to a perceived need to act.

Conversely, overall pressure on politicians to take action on global warming dissipated after 1990, coinciding with a decline in media coverage of the issue from 1990 onwards, despite flurries of attention to key events such as the Kyoto meeting.[7] The decline of global warming as a high-profile political issue and its demise as a media issue are perhaps not unrelated. Declining media attention can be argued to have allowed the salience of global warming as a political issue to wane.[8] As Ungar (1992:494) notes, 'Since the scare subsided, politicians have betrayed their promises.' Many states have retreated from their earlier commitment to reduce greenhouse gases as implementation problems have come to the fore and pressure to persevere has subsided.[9] As established in work that finds correlations between media issue hierarchies and those of the public (Downs 1972), the pressure on governments to act on an issue drops as the subject drifts away from the forefront of public concern.

The media can shorten the time between the passage of an issue from scientific concern to political issue by 'performing an important transmitter role' (Beuermann and Jaeger 1996:191). As Hannigan (1995:58) notes, 'In moving environmental problems from conditions to issues to policy concerns, media visibility is crucial. Without media coverage it is unlikely that an erstwhile problem will either enter into the arena of public discourse or become part of the political process.' The case could be made that it was the media that helped to establish global warming as a political issue in the first place. Beuermann and Jaeger (1996:221) note that the 'obstacles to political agenda-setting were overcome by focusing and raising public

---

[7] For example coverage of global warming in the US press in 1991 was about half the amount in 1990 (Ungar 1992).

[8] As Paterson notes, 'The fact that the media's coverage has declined quantitatively and has become more conservative qualitatively, is a reasonable indicator of declining political pressure on decision-makers' (Paterson 1993a:188, note 7).

[9] See for example Bach (1995) on Germany's inability to meet its earlier greenhouse gas reduction pledge, and Climate Network Europe/CAN-US (1996) for a review of the OECD countries policy measures on climate change, which shows widespread failure to meet the stabilisation targets.

attention by repeated media coverage'. They note how, in the German case, 'alarmed public opinion stimulated by press reports . . . probably played the most important role in shifting the climate change issue onto the political agenda' (ibid.). Ungar (1992:483) argues that in the West in general it was 'public and media response to the alarming heat and drought in the summer of 1988, and not any significant shift in activities by claim-makers such as politicians and scientists, that turned the greenhouse effect into a household term and accelerated political demands'.[10] Global warming had commanded little attention in public arenas prior to this. The media were quick to speculate that the theory of global warming and the unusual weather experienced in 1988 were connected. Evidence of the direct effect of media presentations of global warming upon public opinion can perhaps be found in the correlation between media certitude in reporting scientific scenarios as predictions rather than possibilities, and public certainty about the threat posed by climate change (Bell 1994; Ungar 1992).

Perceptions of intense public concern were directed into the political arena by the media (Ungar 1992). 'Suddenly in late 1988 the climate change issue was an urgent item of public business throughout the West' (Brenton 1994:166). Politicians were caught off guard, lacking prepared interpretive packages on global warming, and were therefore forced to bow to popular concern. Prime Minister Mulroney of Canada committed Canada to a 20 percent reduction in CO_2 emissions and Mrs Thatcher addressed the UN General Assembly on climate change in the autumn of 1989. In the US, two bills were tabled in the Senate on action to tackle climate change, the Environmental Protection Agency produced a report on its possible impact on the US, and presidential candidate George Bush announced that he would use the 'White House effect' to combat the greenhouse effect. In the European Community the Commission produced its first 'communication' on climate change, laying the foundations for joint community action. The agenda had been set.

A further example of media agenda-setting was the trend amongst the US media in late 1989 and early 1990, in the run-up to the Second World Climate Conference, to launch assaults on the scientific consensus on the greenhouse effect. The media buzzed with reports that scientists, 'minus a few greenhouse radicals', in fact doubted the predictions of warming (Athanasiou 1991:6). On 13 December 1989 the *New York Times* ran a front cover feature on the 'Greenhouse sceptics' and *Forbes* (1989) ran an exposé on 'The global warming panic'. Advice from Richard Lindzen and Jerome Namais to the White House that global warming forecasts were 'so inaccurate and fraught with uncertainty as to be useless to

---

[10] Ungar (1992:491) notes, in the context of the 'Greenhouse summer' of 1988, that 'The media served to generalise personal experience. People learned that local, extreme weather was a national, indeed international, phenomenon.' The years 1987 and 1988, along with 1981 and 1983, were among the hottest of the century. The end of 1987 had seen a freak hurricane in the English Channel, and a chunk of ice 25 miles by 99 miles broke off the coast of Antarctica. The following year, 1988, brought hurricane 'Gilbert' to the Caribbean and a catastrophic drought to the American Midwest.

policy-makers' (quoted in Athanasiou 1991:6) were extensively reported across the media. Consequently, when President Bush addressed the IPCC in February 1990 he was able to change his address from one that emphasised the seriousness of global warming to one that emphasised scientific uncertainties, and made no recommendations for action.

Conversely, media outcry at the failure of the White House to support the idea of an international climate change treaty and at the administration's intervention to alter the congressional testimony of James Hansen,[11] contributed to a policy reversal, whereby the US accepted the idea of a treaty and offered to host the first round of negotiations (Brenton 1994:170). This example illustrates the operation of two processes. Firstly, it shows how the media exposé of the Hansen incident created a credibility gap that the US government felt the need to redress by announcing a series of reactive measures. Secondly, it illustrates how the media were able to nurture public expectations about the sorts of institutional policy response considered necessary. By convincing the public that a treaty was a desirable policy move, the media made it hard for the US government to go against the tide of public sentiment. The way in which the US media thrived on the criticism that diplomats and NGOs were willing to offer of the US's intransigent position on climate change (Brenton 1994), also draws attention to the ability of the media to magnify the political impact of pressures exerted by other states upon 'laggards' (Porter and Brown 1991).

Similarly the media criticism of US climate policy that followed the Noordwijk conference in 1989 was responded to by the US government with the announcement of $1 billion for climate research, a plan to plant a billion trees a year, and an agreement to host a White House conference on climate change in the spring of 1990. Domestic criticism championed by and directed through the media was dragging US climate policy into line with that of the rest of the international community.[12] The head of the Environmental Protection Agency, William Reilly, in recognition of the fact that policy was evolving at the pace set by the media, commented, 'We need to develop a system for taking action that isn't based on responding to the nightly news. What we have had in the US is environmental agenda-setting by episodic panic' (quoted in Brenton 1994:243).

Room for manoeuvre in negotiating the climate convention was opened up by a softening of the US's position as the Rio deadline drew closer in mid-1992. By focusing public attention on the summit, in what was an election year for the US president, the media played a part in making more flexible what had been a steadfastly

---

[11] In May 1989 the White House intervened to dilute the congressional testimony of James Hansen, a NASA (National Aeronautics and Space Administration) scientist, who drew attention to the severity of the climate change problem (Brenton 1994).

[12] In the case of Canadian press coverage of global warming, Einsiedel and Coughlan (1993:140) notes that the rise of the issue from 1987 was successful in 'generating its own public and institutional sets of actions and reactions'.

rigid negotiating position.[13] Moreover it is possible to put forward the case that the Framework Convention on Climate Change (FCCC) was *only* achieved in time for signature at Rio because governments sensed that the public expected a convention to be signed in June 1992. The anticipation of a public relations fiasco if this basic demand was not met, helped to force the pace of agreement and injected a sense of urgency into the negotiation process. As Weizsäcker (1994:169) notes, 'Without that momentum and public pressure, it would have been easy for some governments to leave everything to diplomatic routine and thereby let the conference fail.' Given the scientific uncertainties about global warming, the fact the issue touches on key industrial processes and the force of lobbying against mitigative action (see Chapter 5), the mere fifteen months that it took to negotiate the FCCC bears testimony to the importance of the perception among policy-makers that public opinion demanded a policy response. The media, it seems, were instrumental in generating such a perception.[14]

It might convincingly be argued that the media can bring direct pressure to bear most effectively when there is a clear choice to be made and perceptions of popular pressure can be intensified. This fits in with Cracknell's (1993) distinction between 'focused' and 'diffuse' pressure. In the case of the former, media exposure can be 'particularly effective when the issue concerns a clear-cut decision' (Anderson 1997:41). The decision for states over whether or not to be party to the convention seems to have been swayed by such pressure (as the case of President Bush bears out above) and it is possible that future decisions about whether or not to sign up to protocols will be subject to similar pressure. The point is not to argue that the scale of media coverage and its subsequent impact on popular opinion wholly explain policy shifts, since other factors are evidently at play, but to note instead that intensive attention to an issue at key moments, and the degree of public concern this is thought to represent, can contribute to policy movements. At minimum, as Weizsäcker (1994) argues in relation to the convention, the absence of media-generated pressure would have made policy failure more likely.

The influence of the media derives from its access to public constituencies that matter to many governments. This is the 'bargaining asset' that confers upon the media the ability to generate expectations about the need for policy responses. The limits of direct agenda-setting are, however, considerable. It is susceptible to political reassurances, tokenism and incremental organisational change. It is in the private and relatively closed world of government and institutional bureaucracy that issues come, or fail to come, to final recognition through agreement on action. Beyond establishing the issue as one deserving of political attention and generating momentum for a response, the direct influence of the media upon policy may be minimal.

---

[13] As Brenton (1994:191) notes 'In the US President Bush's staff were acutely conscious of the political price (in an election year) of wrecking Rio.'

[14] Jaeger and O'Riordan (1996:18) also argue that media attention to the 1988 Toronto conference on the changing atmosphere had an important effect on the outcome, which was to stimulate action in a number of countries. See also Rowlands (1995).

Moreover studies show that once an issue receives government attention, media coverage declines as a solution is perceived to be in hand (Cracknell 1993). Anderson (1997:42) notes that 'the news media tend to lose interest once an issue has become sucked into the bureaucratic process'. Coverage of environmental issues may also have a greater impact on policy-makers than the public (Anderson 1997), reinforcing the argument developed here, that it is often the perception of the need to act that media coverage can generate that makes a policy impact, rather than an 'actual' body of public concern.

From direct to indirect

Aside from direct media pressure on governments to address an issue, it is also possible to talk more generally of a media *discourse* on the environment that resonates in the public realm. The multitude of ways in which most people read the same text are often situated within the boundaries of broadly conventional interpretations, relating to the fact that alternatives are denied column and air space and therefore remain largely inaccessible. The concept of *polysemy*, which seeks to explain the fact that there can be several readings of the same text, has become entrenched in much audience reception work in media studies (Corner and Richardson 1993; Jensen 1992; Moores 1992) and is useful in developing the idea of a media-constructed frame being more significant politically than the direct impacts described above, by highlighting issues of legitimation and reinforcement. Content retains its importance, but a recognition that more than information is conveyed in news coverage is appropriate. Frameworks, ideologies, narratives and symbols are all contained in news coverage and can be argued to have a significant impact, on the basis of their relative invisibility to audiences. The shift is partly also one of the timing of impact. Direct effects from the provision of new information can be seen almost immediately, as the examples above illustrate. The sorts of effect that are related to the frame of coverage, however, are more long term and depend on factors such as the consistency and repetition over time of the assumptions that underlie the frame.

The need to explore the indirect effects of media coverage is heightened by research that shows a low audience recall of information received through the media, with only about a quarter to a third of basic information being remembered, even after prompting (Anderson 1997; Bell 1994).[15] The details of much of the coverage do not seem to be retained, yet the images they provide or reinforce, the thrust of what is said and the overall message do seem to have an effect on public attitudes. In order to broaden and strengthen this account of the effect of the media upon the scope of climate policy at the international level, these less visible influences are discussed below.

---

[15] Studies carried out in Sweden, the US and Austria produced this figure (Bell 1994).

Indirect effects

The *frame* refers to the choice of topics for coverage, the framework of assumptions, the set of presuppositions within which the issue(s) are presented, and the emphasis and tone: the process of encoding and making intelligible. Frames act as filters to make sense of the world, as representations of reality they can structure the formation of interests and preferences in political debate. Hence explanations of outcomes require some attempt to assess how competing discourses within different formal and informal political arenas can produce (in)action. As Jachtenfuchs (1996:46) notes, such analysis 'asks how frames structure problem definitions and thus open up or prevent possibilities for action'.

What is important here is the selection of certain accounts and interpretations of what global warming means as an issue. Particular definitions are privileged in the organisation of an issue's meaning. Preferred or 'proffered' readings are encouraged by the presentation of the issue before the public, based on a limited ideological and explanatory repertoire. The focusing of attention on one area results in what Bachrach and Baratz (1962, 1970) refer to as 'non-decision-making scenarios' in other areas. Consequently some interpretations of the problem do not acquire salience in public discourse because they are not articulated in public fora. Through this process, more unconventional and challenging accounts are rendered invisible (Golding 1981), and those of *primary definers* in society are upheld.[16] The practice of including and excluding particular ways of interpreting an issue help to establish the boundaries within which public understanding of global warming as a political and social concern takes place.

McQuail (1977) identifies variables that can influence the impact of media coverage upon audiences. These include whether the issue covered is subject to prior definitions, whether the audience has first-hand experience of the problem, the degree to which there are competing sources of information, the extent of repetition and the authority of the sources used. The effect of the media frame used to explain global warming is strengthened by the fact that it has been treated by governments and framed by the media in much the same way as other environmental issues, as an isolated problem not resulting from the prevailing system of political and economic organisation. Moreover global warming is an issue about which the public knew very little before the late 1980s, and was therefore particularly susceptible to media-led constructions. Very few people have experienced at first hand the effects of global warming, the treatment of the issue across the media has shown tremendous similarity, and the complexity of global warming means that it is heavily dependent on specialist interpretations by those groups upon which the media confer legitimacy as authoritative sources. All these factors work to strengthen the impact of media accounts of the problem.

---

[16] Hansen (1993a:xviii) defines primary definers the following way: 'those individuals and institutional representatives who are accessed in media coverage and who help frame and define not only what the issues are but also and more importantly, the terms of reference for their discussion'.

4.5 Media constructions of global warming

This section combines analysis of the nature of media coverage of global warming in a number of states that have been prominent in the international climate change debate, with comment upon its repercussions for climate politics. The constructions described feature the most frequently repeated media accounts of the issue, and are not intended to insinuate that there are not a plethora of different meanings that can be gleaned from the coverage described. Contestation is a central part of societal communication, and there is no single overarching discourse on climate change (Hannigan 1995). The purpose here is to outline the 'proffered' readings that exist in the media on global warming, interpretations that are entrenched in news coverage and resonate both within societal understandings and policy responses.

Despite a previous history of media coverage of environmental issues, it was not until the late 1980s that *global* environmental threats made their media debut, among them global warming. The environment first became a subject of media coverage in the 1960s (Hansen 1991). Following a decline in coverage in the mid-1980s, there was a relatively dramatic increase in media interest in environmental issues during the late 1980s and 1990. Since 1990 there has been a general decline in the volume of environmental coverage in the media, and global warming has not been exempt from this trend (Hansen 1993c). The decline has been hastened since 1992 by a feeling of what Anderson and Gaber (1993b) refers to as 'antisappoint-ment': a negative reaction to the disappointing achievements of the Rio conference.

Whilst the emphasis of the coverage has changed over time and the nature of that coverage varies across media, the content on the whole has shown a continuum. A few basic features have remained constant since the issue of global warming rose to salience in the late 1980s, and these are the focus of the analysis below. Four noticeable trends are used to ground the analysis, with particular emphasis upon the first because it is the most widespread.

The scientific dimension

Early coverage of the problem suggests that the media did not question the science underpinning the threat of global warming (see the section on impacts below). From 1990 onwards, however, there developed a fashion in the Western media to attack the global warming hypothesis. Headlines such as 'More hot air than facts on global warming' (*Sunday Telegraph* 1994), 'The lie of global warming' (*Daily Telegraph* 1991), 'Global warming is a load of hot air' (*Sunday Express* 1991) 'Theories on Ice' (*The Times Magazine* 1993) and 'Experts got it wrong over global warming' (*Daily Express* 1991) illustrate this trend. The trend has been notable beyond the conservative press in the UK, and stories that seek to challenge the theory of global warming continue to be popular. *The Economist* (1998) had a full page article headed 'Global warming: In flux', and *The Observer* (1998) carried a front-page item headed 'Man "not to blame" for global warming', half a page inside on 'Solar wind blows away theories' and an editorial comment, 'So much hot air'.

One way in which the science of climate change has been 'humanised' for audiences is through a combination of sardonic ridicule and an attempt to relate understanding of global climatic processes to personal experience of the weather system. Edwards (1997:4) notes that the 'it's chilly-so-global-warming-is-a-joke' quip has become a perennial feature of media reporting. The poor British summer of 1993 led *The Times* to claim that global warming had been 'revealed as an empty promise'. *The Sunday Times* also attacked those warning of the threat of climate change for 'trying to alarm a sceptical and shivering nation' (Edwards 1997:4). Aside from locating this practice as an attempt to connect local and distant domains, Edwards argues that articles of this type 'albeit benign and humorous in intention, serve corporate goals of ridiculing the threat of global warming' (ibid.).

Another trend highlighted by research on the audience effects of media coverage of global warming, is that a majority of people demonstrate confusion about and ignorance of the causes of ozone depletion and global warming (Bell 1994; Rüdig 1993). As Beuermann and Jaeger (1996:221) note in relation to the German media's coverage of the issue, 'Conspicuously, the issues of ozone depletion and climate change were often mixed up'. Many participants in Bell's research in New Zealand also identified ozone-depleting chemicals, particularly CFCs (chlorofluorocarbons), as being most responsible for global warming. The blending of information about ozone depletion and the greenhouse effect is widespread, with the two issues often being presented as part of the same phenomenon (particularly in earlier coverage), generally with ozone depletion being seen as the cause of the greenhouse effect. In Bell's research, of those people asked about their understanding of global warming, only a quarter cited fossil fuels as a cause of the problem, and fewer still mentioned CO_2 or other greenhouse gases (Bell 1994). Interviews in the US have also shown that 'Americans connect global warming to pollution, but not to energy consumption' (Kempton 1991). Since most people regard CFCs as the primary greenhouse gas, the perception has developed that once CFCs have finally been eliminated, the global warming problem will in effect be resolved. Consumption of fossil fuels is rarely acknowledged to be implicated in the problem. This is an area that any strategy to deal with the emission of greenhouse gases, if it is to be effective, will have to address, and it is also the area that is most prone to political controversy.

Despite the general finding that exact details and factual information are not on the whole retained by the public, factual errors in media coverage have been reflected and duplicated in public misunderstanding. Some cite this confusion to explain why 'the public is ambivalent over policy measures, especially new taxation' (Jaeger and O'Riordan 1996:27). Bell (1994:59) concludes, 'It may well be that the public ignorance of the causes of global warming and their confusion of those causes with ozone depletion have permitted political inaction on this issue.' This is supported by Rüdig (1995), who has found considerable confusion among the public not only about what global warming is, but also, perhaps more significantly, what should be done about it.[17] Anderson (1997:184–5) notes that 'These misunderstandings may have serious consequences for preventive behaviour.' For example interviews show that many people do not make a connection between using energy more efficiently and curbing global warming (Kempton 1991).

Following from the science-based focus on coverage on global warming is what CNN's Tessa Ryan refers to as 'psychic numbness' among a public besieged by conflicting studies, predictions and uninterestingly reported facts (quoted in Studelska and Buchaman 1993). Number-heavy stories dampen public response, and are thought to lead to what has been termed 'information pollution' (Studelska and Buchaman 1993). The underlying message of stories is often lost in a maze of decontextualised studies and reports on the state of the world's climate. Dangers couched in the distant dialogue of science and statistics, lessen the impact of the message contained within.

It can also be argued that the media's inclination to construct binary oppositions creates distortions, in terms of accurate presentation of the degree of scientific consensus that exists on climate change. As far as some commentators are concerned,[18] commonplace media emphasis on uncertainty in the understanding of greenhouse science, 'directly contradicts the scientific consensus and the conclusions of the UN IPCC's scientific committee' (Athanasiou 1991:7). As the environmental digest *ENDS* notes, 'Readers following the greenhouse debate in the media may be forgiven for being confused about the current situation. Doubts, questions and refutations of the theory have been promulgated widely' (*ENDS Report* 1992a:5).[19] It can be argued that media coverage of global warming is not representative of the balance of views within the scientific community. In other words, the attention that some actors receive is disproportionate to the representativeness of the claims they make, despite the media contention that 'The debate over global warming is vigorously two-sided' (Santaniello 1997). Reporters, unwilling or unable to draw conclusions, fall into what is described as the 'balance trap' (Studelska 1993), whereby an opposing viewpoint is sought even if it is an extreme one.[20]

Fred Singer, a prominent greenhouse sceptic who has repeatedly been called upon by the media (particularly in his home country, the US) for his opinions, is

[17] Bostrom (1994b) also shows that audiences have a poor appreciation of the fact that global warming will result from a rise in atmospheric CO_2 and that the single most important source of CO_2 is fossil fuels. Bostrom also reveals that audiences tend to believe that global warming has already occurred, and that there is considerable misunderstanding about the causes and effects.

[18] *ECO* notes that 'the recent spate of polyamish articles in the popular press, assuring us that the scientists' predictions of substantial negative impacts from greenhouse warming are greatly exaggerated, are based more on a handful of highly vocal "greenhouse sceptics" than they are on the broad mainstream of climate scientists' (*ECO*, issue 3, Geneva 1993).

[19] NGOs have expressed concern that press reports have 'given the impression to many people that the threat of climate change may not be as serious as was previously believed' (*ECO*, issue 6, Geneva 1994).

[20] Hannigan (1995:68) refers to this as the 'equal time' technique, where rival claim makers are quoted out of context.

an example of a marginal figure in the debate being conferred vast access to influential media.[21] Another familiar sceptic, Patrick Michaels, who publishes a newsletter funded by coal interests, 'is cited in the media almost as much as Stephen Schneider of Stanford, a pioneering climate change researcher', reflecting, according to Allen (1997), 'American journalist's tendency to accentuate extremes in their effort to get both sides of the story.' Scientists who argue that temperature variations can be attributed to sunspots and variations in the sun's output 'have received more of an ear in the popular press than in the scientific community' (*ENDS* 1994:22). Greenslade (1996), commenting on coverage of global warming in the UK press, notes that 'Every right-wing paper has attempted to debunk global warming, providing space to "experts" who happen to share such views.' Moreover, when fossil fuel lobby groups accused two lead authors of the IPCC's *Second Assessment Report* (1995) of deleting passages expressing uncertainty about the climate threat, their claims were reported widely. The *Wall Street Journal* (1996) spoke of 'A major deception', and the *New York Times* (1996) declared that the report had been 'improperly altered'.

The effect of such coverage may be to encourage the view that no scientific consensus exists on global warming; to position global warming as a hypothesis rather than a fact, even though the IPCC (1995) has declared there to be a 'discernible human influence on the climate system'. This is borne out by Rüdig (1993) and Bell (1994), who illustrate that public uncertainty about climate change runs very high. At the moment the coverage, constructed as existing between two polarities (believers and sceptics), swings towards those who reject the global warming hypothesis.[22]

Impacts

Consistent with the observation that the media tend to focus on the more dramatic aspects of the climate change issue, is a notable fascination with the predicted impacts of climate change. Initially the issue of global warming attracted a great deal of sensationalist coverage, riding as it did on the back of media fascination with ecological threats of enormous magnitude. Whilst there seems to be little evidence of deliberate distortion on the part of the media, there is evidence of the media treating as fact scientific *estimations*, and taking worst-case scenarios as probable. Huge sea-level rises were mentioned (30–60 metres) as the possible outcome if the polar ice caps were to melt, without mention of the fact that this would take hundreds of years (Bell 1994). Doomsday scenarios are constructed and impacts made to sound near-term, supporting the notion that one of the key functions performed by the

---

[21] See for example, 'No scientific consensus on greenhouse warming' (Singer 1991) and 'Earth Summit will shackle the planet not save it' (Singer 1992a).

[22] Professor Bert Bolin, chairman of the IPCC, was moved to warn delegates not to be confused by reports in the press. Bolin described media reports on global cooling as exaggerated and inaccurate (*ECO*, issue 1, Geneva 1994).

media in environmental debates is to connect distant and local domains as well as couch long-term time frames in terms of the immediate (Burgess et al. 1991). In 1988 the media took it as given that global warming was an immediate and serious threat. Media overstatement about climate change was associated with public overestimation of the principal effects of climate change (Bell 1994; Ungar 1992), further emphasising the media's agenda-setting function.[23] There is some evidence of this trend continuing in certain media.[24] In an article titled 'Meltdown', Brown (1996b) declares that 'Global warming is no longer tomorrow's worry. High up in the Swiss Alps, you can actually see it happening.' In general terms, however, Ungar (1992) notes that the more alarmist edge to media coverage of global warming has been tempered and predictions and prescriptions have become less radical, coinciding with the gradual consolidation of a more sceptical position amongst certain sections of the scientific community with regard to the global warming hypothesis (see above).

As will be discussed below, the mass media prefer 'event-oriented' coverage of environmental issues; dramatic stories make good copy because they accord with a fairly clearly defined set of news values. Occasionally on this basis, climate change (when reporters are willing or able to forge the connections) draws attention to itself. The story that landslides hitting the US may be 'the first tangible consequences of global warming' is a case in point (Irwin and Davies 1997:19). Often stories are run that relate global warming to 'softer' ecological or conservation-oriented concerns such as wildlife protection. Such stories provide the opportunity graphically to represent pictures of affected species, for example, which enable audiences to visualise the possible impacts of climate change.[25] Making climate change more palpable serves to amplify public perceptions of risk by bringing them closer to home (Anderson 1997:187–8).

In terms of generating policy demands, this sort of coverage is likely to engage existing public concern. As Chapman et al. (1997:76) argue, 'it is not abstract things like General Circulation Models . . . which have stirred the government and some people to action; it is real problems in the country'. The extent to which this is the case clearly differs from country to country. Comparing the UK and India,

---

[23] Professor Bert Bolin, chairman of the IPCC, was alert to this trend when he noted that 'Journalists put about wild exaggerations which naturally scare people. There may be an utterly small chance that things turn out in such a way. We don't need to scare people, instead we must bring home to them that we face a serious issue in the future development of our world and set in train a process of collaborative action' (*ECO*, issue 7, Geneva 1990).

[24] See for example 'Beaches to vanish as sea rises 3ft' (Worthington 1992), 'Millions at risk in expanding dust bowl' (Schoon 1994c), 'Melting Antarctic sounds alarm for globe' (Schoon 1994d), 'Melting polar ice will flood coastal towns in 40 yrs' (Smith 1992), 'Global warming could bring disaster to low ports' (Sheppard 1992), 'Going the way of Atlantis' (Vidal 1992), 'Rising seas may drown Hong Kong, Tokyo, Rio' (Ghazi 1992), 'Millions could go hungry due to global warming' (Schoon 1992), 'The world faces series of disasters' (*Evening Standard* 1992b), 'Warming may bring malaria to Britain' (*Evening Standard* 1992a). The *Florida Sun-Sentinel* declared recently that the melting of the polar ice caps will mean that 'seas could rise 220 feet' (Santaniello 1997:1).

[25] See for example 'Global warming threat to 40% of Britain's birds' (McIlroy 1998:7) and 'Global warming damages part of Barrier Reef' (Rennie 1998).

Chapman et al. note that whereas in the UK scientists have been the primary definers in the public debate on climate change, in India 'it is local trauma such as drought and famine and large-scale displacement of people which are more likely to be triggers of heightened attention' (ibid.:88). This supports the contention of Burgess and Harrison (1993) that people mediate global issues through local lenses, 'watching globally, thinking locally' (Chapman et al. 1997:169). Just as through initial emphasis on impending disaster, global warming became an issue of crisis management rather than structural reform, so focus upon the calamity of these 'natural' events overlooks the indictment of the patterns of behaviour that give rise to them.

The economic dimension

Throughout the history of media coverage of global warming, excessive costs have frequently been the focus of reporting emphasis.[26] Despite evidence that energy efficiency and the alternative energy paths needed to combat global warming could ultimately save money (Jochem and Hohmeyer 1992; Lovins and Lovins 1992; Mintzer 1992) the media regularly portray action to combat global warming as so costly as to retard economic growth severely (Ungar 1992). In the seven years following the summer of 1988, all media sources investigated by researchers have featured stories and editorials stressing costs and the need for a 'cool-headed response' (ibid.). An *Economist* cover story 'Warm World, Cool Heads' offers a clear articulation of this line (*Economist* 1990:13). The same publication later went on to comment that 'One of the few certainties about global warming is that the costs of severely curbing emissions of greenhouse gases now would be huge' (*Economist* 1995:11).

Often apparently 'objective' and authoritative sources are drawn upon to assess the costs of climate action. The *Wall Street Journal Europe* called upon an economic adviser to the Norwegian Conservative Party, Australian Prime Minister John Howard (known for his hostility towards action on climate change) and the Bureau of Resource Economics (with close ties to the Australian coal lobby) to piece together its case that 'Rio set the bar too high' (Bate 1997). Studies produced by conservative think-tanks on the cost of climate mitigation action 'have been given solemn coverage by the press' (Ivins 1998). Through TV slots and press reports of these studies, 'Americans are warned about a broad decline of US competitiveness, reduced living standards, and the migration of industries to developing countries' (Krause 1997:1). The effect of this coverage has been to influence 'a significant fraction of the US political elite' (ibid.). Athanasiou (1991:3) argues more generally that the radical and uncomfortable message behind greenhouse warming scenarios is smothered by the mass media with its emphasis on 'cost and the putative dangers of over-reaction'.

[26] See for example 'Energy industry warns of heavy cost' (*Financial Times* 1993), 'Pollution charges put luxury cars at risk' (*The Times* 1993), '10 pc rise in fuel prices plan to cut carbon pollution' (*Daily Telegraph* 1992), 'Carbon tax no use say greenhouse sceptics' (*Daily Telegraph* 1994a).

Declining attention to the international

A further trend that has emerged in media coverage since 1992 is a general decrease in attention to developments in the ongoing international climate negotiations (Anderson and Gaber 1993b; O'Riordan and Jaeger 1996), from the high degree of coverage given to early international meetings to address the issue. At the 1988 Toronto Conference on the Changing Atmosphere, for example, extra press rooms were required to accommodate the huge influx of international journalists (Ungar 1992). The trend towards declining interest at the international level has been particularly noticeable since the Rio conference, where the politics of climate change took on a new importance and required large amounts of pressure to push the negotiation process towards a binding protocol. This can be partly explained by the post-Rio 'antisappointment' (Anderson and Gaber 1993b) referred to above, but also by the fact that negotiations are rarely seen as newsworthy given their prolonged, monotonous and relatively staid nature. Very little of dramatic consequence appears to be achieved at any meeting of the Conference of the Parties, and the costs of attendance can be astronomical and difficult to justify against a background of cost-cutting.[27]

To some extent the trend in media coverage away from events in international fora leaves governments unaccountable for the decisions they take, given that news of these decisions is increasingly not disseminated to domestic public audiences. Where there is mention of negotiations, it is brief, highly factual and superficial in terms of probing the issues that underlie the negotiations. This trend is borne out by the coverage on the first Conference of the Parties in Berlin in spring 1995. Not one of the mass-market newspapers in the UK covered the conference while the *Financial Times* and the *Guardian* only addressed the conference itself, rather than global warming as a general news issue (Russel-Jones 1995). BBC TV news in the UK (at six and nine o'clock in the evening) used the conference to introduce a speculative piece about whether the snowfall that March was evidence of a changed weather pattern attributable to global warming.

The Kyoto meeting (COP3) in December 1997 attracted the most media attention since Rio, because a tangible outcome seemed likely and all the essential ingredients of a good story were there: the conflict between the US and the EU,[28] the last-minute backroom deals, and the introduction of new issues into the debate such as 'hot air' trading and permit trading.[29] Where there is conflict, international meetings will continue to attract some coverage.[30] It is also possible that the scale of media attention to the Kyoto meeting acted as a catalyst for a decisive outcome.

---

[27] See Anderson and Gaber (1993a) on the way in which the imperative of cost-cutting post-Rio has affected coverage of international negotiations.

[28] 'Clinton on course for a clash with Europe over global warming' (Clover 1998:14).

[29] 'Developed nations look to a free market in pollution' (Schoon 1997:5).

[30] 'Japan attacked as climate deal nears' (Boulton 1997:4), 'Pollution treaty in the balance after rebellion' (Clover 1997:20), 'Anger at deal loopholes as climate talks drag on' (Brown 1997:3).

Rather like Rio, potentially very diffuse sources of pressure are felt more acutely when there is a clear-cut decision to be made (to sign or not to sign) and the international media spotlight is focused squarely on the final agreement of the protocol.

Aside from the highpoints of the climate negotiations however, when deals are struck and months of laborious negotiations result in an agreement, media coverage of the negotiations nomally amounts to little more than passing interest. A quote from journalist Nick Nuttal sums up the situation: 'Unless there's a cracking negotiation going on you're really bashing you're head against a brick wall as far as getting the piece published in the paper is concerned. Negotiations are going on all the time . . . there is fundamentally nothing new to say.'[31]

The up-shot of these various trends is that public concern about global warming is generally low, and on the basis of this recalcitrant governments have felt able to emphasise the absence of a mandate to act.[32]

4.6   Explanations and consequences

This section examines three approaches that are brought together to help account for the nature of this coverage, and therefore explain the nature of the constructions of global warming described above.

There may be one all-embracing explanation for the nature of media coverage of global warming, such as the 'issue attention cycle' (Downs 1972, 1991). This term describes a process in which issues have a political life that begins with initial concern and alarm, but becomes tempered as factors such as cost are brought to bear and the difficulties of action become apparent, whereafter public interest moves (or is led) elsewhere, often leaving the original problem unresolved. Downs predicts, however, that environmental problems are unlikely to fade from media attention because they tend to be visible, solvable by technological means and can be attributed to the activities of a small group in society. The problem is that none of these characteristics accurately depicts the complexity and challenging nature of the climate change problem.

The purpose of this section is to gain a better understanding of the nature of the media agenda on global warming by looking at different influences on the formation of that agenda. Three main approaches are briefly discussed, based loosely on predominant schools of thought in the media studies literature (Curran et al. 1982).

The political economy approach

This approach to understanding media coverage focuses on questions of ownership and corporate finance and the effect these may have upon news content.[33] Private

---

[31] Interview with Nick Nuttal, 25 February 1994.

[32] See for example *ECO*, issue 4, Geneva 1994, and *ECO*, issue 2, New York 1994.

[33] See Herman and Chomsky (1988), Curran (1988) and Garnham (1986) for articulations of the political economy approach.

corporations are particularly significant in this respect, but it should also be recalled that governments are often the single biggest source of advertising revenue (McQuail 1993:134). At the most basic level, the approach builds on the idea that the media are constrained by the hand that feeds them. On a more subtle level it looks at the internalisation of investors' interests by journalists, which serves to marginalise discussion of certain issues or bounds the way in which those issues get discussed. Hannigan (1995:67) argues that reporters' sensitivity to 'external pressures from corporate advertisers' means they may, on occasion, 'modify or deliberately overlook significant stories which involve environmental wrong-doing'. Tom Athanasiou (1991:1) argues in relation to media coverage of global warming, that 'The frames within which it is presented reveal with rare clarity the presumptions and prejudices of the framers.' For him, the framers are the media owners and investors; for the most part corporations.[34]

'No-go' areas of critical reporting (Golding 1981) would undoubtedly extend to an issue such as global warming, given that the media are increasingly linked to companies operating in areas such as oil (Golding 1981:140; Hollingsworth 1986). Since the media are controlled by parties with a direct interest in the global warming issue, their assessment of suitable courses of action and the effectiveness of prevailing government strategies is arguably loaded. In instances when the media have been critical of governments' inaction over global warming, political economy accounts would explain this in terms of the different corporate interests that exist in relation to global warming or the absence of a strong corporate interest in the issue. Nuclear energy interests, which are using the issue of global warming for self-promotion, are also investors in the media and therefore coverage has been forced to tread a delicate line between different corporate interests.[35] The political effect is to compromise the independence of news coverage of issues where sponsors have vested interests, and to ensure that public understanding of the problem is not damaging to their business.[36] In an honest appraisal of the situation, the manager of corporate communications at General Electric in the US said of the utility's control over NBC and ABC news, 'we insist on a program environment that reinforces our corporate messages' (quoted in Herman and Chomsky 1988:16).[37]

---

[34] Chomsky notes that of the 1,800 daily newspapers, 11,000 radio stations and 2,000 television stations in North America, twenty-three corporations own and control over 50 percent of the business in each media, in some cases enjoying virtual monopoly (Achbar 1994:62). For further details about increasing concentration of ownership see Golding (1981) and Bagdikian (1987).

[35] Indeed Anderson (1997:66) documents the case of a journalist who claimed considerable editorial interference on the ground that the stories he was writing were critical of the nuclear establishment.

[36] Ian Breach, a BBC TV environment correspondent and former journalist with the *Guardian*, observed, 'a pervading fear that valuable advertising will drain away in the face of persistent criticism' (quoted in Curran 1986:331).

[37] This input is often organised through organisations like AIM (Accuracy in the Media) which are financed largely by corporations to put pressure on the media to follow their preferred agenda. It is also interesting to note that eight separate oil companies were contributors to AIM in the early 1980s (Achbar 1994:57).

The exercise of a veto over coverage does not have to be direct, however. It is often the case that journalists and their editors engage in self-censorship in anticipation of intervention when they cover particular issues in a certain way.[38] As Golding (1981:140) notes, 'the fact that allocative controllers may not intervene in routine operations on a regular basis does not mean that there is no relationship between the owners' ideological interests and what gets produced'. Tacit pressures derive from a dependency on revenue from advertisers and are manifested in what is included and excluded and how the issue is couched. As Curran et al. (1982:18) note, 'The workings of these controls are not easy to demonstrate or to examine empirically. The evidence is quite often circumstantial and is derived from the "fit" between the ideology implicit in the message and the interests of those in control.' Ties with fossil-fuel-dependent industries may explain why, even though the intensive use of fossil fuels is implicated in the global warming problem, media reports tend to cloud the issue by encoding the news in terms that emphasise scientific remedies, technological fixes or simply the extraordinary nature of the issue. Awareness of the interests of key investors may account for the generally very negative coverage of proposals for a EU carbon tax in much of the media, and repeated emphasis on the costs the proposal would inflict upon industry (see Section 4.5 above). As Andrew Veitch, environment correspondent at Channel 4 News, notes, 'When green issues came to be seen as a threat to financial interests, then the enthusiasm declined suddenly . . . as soon as the environment correspondents . . . started looking for the causes of things . . . editors became rather less keen on running these stories' (quoted in Chapman et al. 1997:44). Because the solutions to global warming have the potential to require significant economic restructuring, coverage has tended to depoliticise the issue. This would account for the preference for coverage in which 'the environment is taken to be a politically-neutral area relating to the quality of life rather than its organisation' (Lowe and Morrison 1984:88).

This relates to Edwards' (1997:2) account of the media's coverage of global warming: 'the perceived seriousness of a threat is largely determined by the extent to which it is a help or a hindrance to goals set by centres of political and economic power'. He notes further, 'The media are thus unlikely, even structurally incapable . . . of offering root cause analyses of the problems we face today', interpreting environmental problems as isolated incidents rather than symptoms of broader economic and social practices.

Social dimensions

This approach to explaining the nature of the coverage relates to the professional norms of journalists and the 'organisational routines within the news room'

---

[38] Schlesinger (1992:296) notes that, 'on the whole the subordination of the media within the wider structures of power ensures that the system works perfectly well without any need for open intervention or censorship'. Chomsky argues that the media have 'learned to be sympathetic to the most delicate sympathies of corporations' (quoted in Achbar 1994:63).

(Hannigan 1995:59). It is important to note, firstly, that the issues the media choose to cover do not necessarily correlate with the status of the issue as a pressing concern (Downs 1972, 1991). As Lacey and Longman (1993:227) observe, 'Despite the gravity of the predictions which followed the acceptance of the greenhouse effect/global warming hypothesis by IPCC ... the number of articles written on these topics declines rapidly in the Autumn of 1990 and the Spring of 1991.' Following the publication of the IPCC's *First Assessment Report* in 1990, the *New York Times* ran the headline 'US data fail to show warming trend', while *Forbes* magazine heralded 'The global warming panic: A classic case of over-reaction' (quoted in Edwards 1997). Similarly, ten days after the IPCC reported on 6 June 1996 that 'the balance of evidence suggests a discernible influence on global climate', *The Sunday Times* declared that 'The latest apocalypse, global warming, is just that. Lots of hot air' (ibid. Edwards 1997:4), while the *Daily Telegraph* offered the view that 'to many scientists, the likelihood of man-made global warming is about as credible as stories of goblins and fairies' (Edwards 1997:4).

This phenomenon relates not just to the scientifically determined importance of the issue. When an issue takes on political significance it is not necessarily accorded coverage either, as issues compete for media attention. For example, by May 1991, as negotiations on the convention got under way, media interest in global warming had waned. More recently, despite the warning by UK secretary of state for the environment John Gummer (during COP2), that 'Global climate change needs action now' and that 'alarm bells ought to be ringing in every capital of the world', 'Those papers which greeted the conference by accepting its central thesis assumed they had done enough. Papers which cannot stomach the scientific evidence for global warming ignored it' (Greenslade 1996). Greenslade concludes, 'This latter attitude leaves readers seriously uninformed about a serious issue' (ibid.).

Media coverage tends to follow trends, adapting to what is deemed 'fashionable' at any particular time.[39] For example, journalists describe global warming as having 'gone off the boil a bit'.[40] The media seek out the short-term, novel and manageable aspects of current issues. The issue of global warming, therefore, does not sit easily with traditional news values and representational practices. The media desire simplicity where in global warming they find only complexity and uncertainty. As one journalist said of global warming: 'There's no clear message emerging ... it's not as clear-cut as it was a few years ago' (Schoon 1994a). There is nothing concrete that can be reported – speculation is rife, but hard news on the issue is scarce. Global warming is less easily adapted to familiar narratives about 'goodies' and 'baddies'. Instead, responsibility is more diffuse, and the causes more ingrained in everyday behaviour.

---

[39] This is illustrated by BBC TV and Radio environment editor Alex Kirby's remark that 'There's a tendency for things to go in fashions ... we as journalists, correspondents and editors think we know what the public want ... we say "Oh no, the environment's not an issue any more, it's old hat" and we move on' (interview with Kirby, 25 February 1994).
[40] Interview with Nick Schoon, 25 February 1994.

A fundamental difficulty is that while environmental degradation can be a drawn out process that is not clearly visible, the media feed on short, sharp, highly visible events. As Lawrence McGinty puts it, the issue is 'Can we show the threat, can we demonstrate that threat in pictures. If we can't, well, forget it' (quoted in Chapman et al. 1997:47). Global warming has also entered that stage of the 'issue attention cycle' where its technical and political complexity has become more apparent, where-upon media attention often declines. The continual search for a new line on the familiar story of global warming helps to explain why 'repeated stories about climate change are likely to be valued less than a controversial challenge to expected predic-tions' (Anderson 1997:119). The long time frame in which climate change operates is occasionally punctuated, however, by 'disasters' such as the drought of 1988 (referred to above), which encourage speculation about global warming as the possi-ble cause and therefore represent a newsworthy event.

On the whole, sudden events such as oil spills receive greater media attention than long-term processes such as global warming. This is accounted for by Chapman (1992:33) by the fact that 'News is based on high-frequency expectation of low-frequency events. If degradation is slow and always with us, then it is not a low-frequency event.' In explaining different types of coverage, a distinction can be drawn between the environment as accident, which by its sudden, unex-pected and visible nature 'forces' itself on to the media agenda, and environmental issues that only become well-known because of the claims someone makes about them. The former will always receive news coverage.[41] In Chapman's (1992) ana-lysis of the content of global television coverage on a single day, nearly all reports on environmental issues had as their subject matter an 'eco-catastrophe', or else the issue was covered in relation to a 'weightier' issue that was regarded as having higher political capital. Environmental issues in all the television coverage were reported as 'events'. This is consistent with the point made earlier that media cov-erage of global warming is event-oriented, it is not contextualised as part of a wider set of political and economic processes. In this way the freak nature of the event becomes the focus of attention, and the issue is understood as 'the recent outcome of an event rather than the inevitable outcome of a series of political and societal decisions' (Hannigan 1995:65). Many of the examples of coverage cited earlier in the chapter, in which the media thrive on the drama and magnitude of potential climate destruction in a way that does not posit connections between human behaviour and the disasters they describe, bear testimony to this trend.

Event reporting also explains in part why coverage of policy responses in interna-tional fora is so scarce. The drawn-out and complex nature of international negotia-tions disinclines them to media attention. The professional criteria of novelty and proximity become increasingly incompatible with protracted negotiations at the

---

[41] The World Coal Institute's climate negotiations bulletin, *ECOAL* (1996:1), argues in this respect that 'A tip falls off an iceberg in the Arctic and the media from around the world are quick to prophesy our approaching doomsday. But reputable scientific work of extreme importance is often neglected and does not see the light of day.'

international level over seemingly intractable issues. Incremental movement of policy positions and tedious redrafting of convention texts is not the stuff of sound-bites and eye-catching headlines. As a conference on the subject noted, 'While the media have proven willing to give some coverage to one-off events such as major conferences, the dramatic implications of such political and scientific statements have not fed through day to day reporting of "mainstream" transport, energy, economics and business stories' (Royal Society 1998). Due to the organisation of newsrooms into distinct issue areas, climate change, as with all other environmental issues, is understood as a discrete problem that has no bearing on other areas of social and political life.

The limited coverage of the international context of global warming is further accounted for in the social dimension approach by the way in which national media tend to rely on their own governments as sources (McQuail 1993:131). Chapman et al. (1997:18) argue there is a structural bias against receiving stories from Southern nationals and agencies even if a developing country is the feature of the story. The preference appears to be for stories from more familiar Northern sources, about the South. As a result the global context, which is central to understanding global warming, is marginalised.[42] McQuail (1993:131) goes as far as to claim: 'Often it seems that the mass media are effectively assimilated into the goals of national foreign policy.' Part of the explanation of the bias towards government sources is that it is often convenient to accept information produced by national institutions for reasons of cost and routine (Gandy 1991). Media interpretations of the global warming problem are equally restricted by their bias towards government information. Hence it is not that the media are necessarily the primary definers of news, but that their relationship with powerful primary definers accords them a crucial role in (re)producing the definitions of those who have privileged access to the media.

More generally, the extensive use of government information can be explained by a media culture that privileges figures of authority over other journalistic sources. The forum of formal political activity is near the top of a media source hierarchy, together with the scientific community. The linkages between these fora and the media are stronger than those between the media and environmental pressure groups for example (Hansen 1991). Questions of credibility, trustworthiness and access to resources conspire to exclude less official sources of information. In an analysis of television coverage of environmental issues it was found that government and industry accounted for 28 percent and 13.2 percent of sources respectively, while advocacy groups accounted for just 6.8 percent (Chapman 1992). Likewise, in a comparative study of Danish and UK television it was found that environmental groups were considered 'primary definers' in only 6 percent of stories (Hansen 1991).

---

[42] Cottle (1993), for example, has found that stories with a global reach have the least of all TV news exposure.

Moreover, while pressure groups can on occasion play a role as claim-makers, it is to the fora of formal politics and science that the media turn for validation of such claims (Hansen 1993b; McQuail 1993:130).[43] The way in which the Greenpeace International report 'The Climate Time Bomb' (1994a) was dismissed as unscientific across the media, via reference to the opinions of climate scientists (unconnected to the report), is just one example of this tendency. The report was assessed for its worth against the opinions of scientists, who were regarded as more authoritative sources.[44]

Hence whilst accommodating different perspectives, the media deal with more controversial points of view by investing establishment interpretations with greater significance and thereby according them the mantle of 'common sense' (Anderson 1997:22). The effect of mediating global warming through 'experts' as the voice of authority on the issue may, as Anderson (ibid.:115) claims, be to discourage both critical thinking and lay views on climate change. Although people tend to view global warming as a serious threat, focus group work on audience reception suggests that global problems are 'never discussed as problems which could either affect them or about which they should or could take substantial action' (Chapman et al. 1997:179). The privileging of science in media discourse serves to abstract the problem of global warming from its origins in the routine practice of economic life and displace it to the global level and expert arenas.

Consequently, future media coverage of global warming as an issue hinges crucially upon the extent to which it becomes part of, and is articulated through, the agendas of these other established fora (Anderson 1993:55). Indeed it might be argued that the rise of concern about global warming owes much to the media projecting the claims-making activity of scientists into public fora (Hansen 1993b:150). This trend has been exacerbated by the high degree of specialisation of those who are cast as experts in the global warming debate. Here the media are particularly dependent on scientific sources and are in less of a position to challenge the information and opinions they receive (Corner 1991), as the prevalence of the 'balance trap' (Studelska and Buchanan 1993) described above bears out. Politically, the system of source hierarchy has the effect of privileging certain types of solution: those solutions articulated in formal political or scientific fora. Media production practices incline towards reinforcement of the paradigm implicit in the constructions described above, characterised by an emphasis on incremental management through techno-scientific solutions.

---

[43] This may explain why environmental groups such as Greenpeace have attempted to nurture an 'alliance with science' to raise their standing as a news source (Hansen 1993b). See Anderson (1993) and Hannigan (1995) on how the ease of access for different sources varies over time.

[44] See for example Schoon (1994b).

The culturalist approach[45]

Less concerned with media production processes and more with the cultural norms embedded in media coverage, is what might be termed the 'culturalist' approach. This approach emphasises the 'cultural givens' (Hansen 1991) of society, which are reflected and reinforced in media coverage. By facilitating and delimiting the nature and scope of media coverage of global warming, these givens have a decisive influence on the terms of debate within which policy strategies will be sought. In order to gain prominence in the public sphere, an issue has to be cast in terms that resonate with existing and widely held cultural concepts. The extent to which the issue becomes legitimised depends on the strength of its relationship to prevailing values (Hannigan 1995; Hofrichter 1990). The importance of the wider cultural setting can be seen in the fact that coverage differs across different societies.[46]

In broader cultural terms, a modernist faith in progress through science and technology explains, to some extent, why the media discourse on global warming is in many ways a science discourse, drawing upon scientists as the primary arbiters between 'right' and 'wrong', 'true' or 'false' (Hansen 1991), and why technoscientific solutions predominate in coverage as appropriate remedies to the problem. One example of the way in which global warming is conceived as a strictly scientific/technical problem by the media, was a piece published in *The Times*, which suggested that 'The problem of global warming is closer to solution thanks to a satellite called ERS-1' (*The Times* 1992). Anderson (1997:128) finds that despite the fact that few journalists have any scientific training, they have a high degree of faith in science. An inability to discern the credibility of scientific reports may explain the uncritical way in which they are often reported, a trend exacerbated by the coverage of environmental issues by reporters from 'non-environment' desks.

Once an authoritative community – as perceived by the media, in this case the scientific community – demonstrates concern about global warming, status is conferred on the issue as one worthy of investigation. A quote from *The Times* environment correspondent, Nick Nuttal illustrates this: 'It's about reporting the growth in understanding in the scientific community about these issues. ... I think that it's a little conceited to believe that you can set your own individual agenda outside that of scientific opinion and evidence.'[47] In other words, global warming is a scientific issue, scientists set the pace and the job of the journalist is to track developments in the scientific community. As shown above, global warming is framed as a distant-public discourse characterised by abstract scientific arguments about climatological processes. The issue comes to be viewed as a strictly technical question, dependent for its resolution on the removal of scientific uncertainties. In this way the question of responsibility and obligation to those societies which are likely to be hardest hit by climate change, or to the future generations that will inherit the problem, is

---

[45] This term is adopted from Anderson (1997:9).

[46] Linne (1993) convincingly shows this in relation to the different nature of coverage in the UK and Denmark.

[47] Interview with Nick Nuttal, 25 February 1994.

marginalised in the debate about appropriate responses by divorcing the issue from its political context.

Taken together, sponsorship activities, professional practices and cultural 'givens' help to explain the nature of the interpretive package deployed by many media in their construction of global warming as a political issue. To some extent, these categories overlap. Governments and corporations react to and are conditioned by cultural and social forces and values. Similarly the social and cultural background is influenced by the activities of governments and corporations.

It appears from the overview of the above approaches that the economic, social and cultural production of news, as it is currently communicated through the media, is juxtaposed with the advancement of an agenda to confront climate change that addresses the fundamental questions of production and consumption, which are central to the problem.

4.7   Conclusions

It seems that the media exert most influence, in a direct sense, over the initial agenda-setting stage of the policy process, when they can help to define the interpretive context of a new issue and create expectations about policy responses from governments. While their influence undoubtedly extends beyond this phase (as the examples concerning the international negotiations illustrate), it is their influence over the framing of an issue for policy attention that has the most political impact. It is the translation of scientific concerns into risk perceptions that enables the media to raise issues that linger in scientific journals to the status of a current political issue deserving of a policy response. By shaping public opinion, a situation can be created where it is conducive for governments to act, or hard for them not to act in the face of perceived pressure to initiate a policy response. Whilst media-generated public interest was sharply focused on global warming, new institutions, programmes and policies were created to address the issue, that persisted long after public attention had shifted elsewhere. It will be harder, however, in the absence of such pressure to maintain the momentum required for states to pursue binding commitments to limit greenhouse gas emissions. Global warming is particularly vulnerable to an issue-attention cycle of this sort. The majority of people are not yet suffering. Global warming is at root a problem generated by arrangements that benefit powerful sectors of society, and few practicable techno-fixes are available. These qualities ill-dispose it to sustained media and public interest.

The greatest influence the media may be said to have upon global climate politics, however, is the privileging of particular interests and perspectives. The frame retains relevance over time, as a context in which future decisions about climate policy will be made, in a way that direct agenda-setting does not. The net effect of media coverage seems to be the reinforcement and consolidation of a perspective located within the existing structures of political and economic power. Exclusion, through the setting of particular agendas at the expense of others, constitutes a form of power

(Bachrach and Baratz 1970). This is the form of power that the media exercises in the politics of climate change: the privileging of some interpretations of the problem and the marginalisation of others.

A mere upswing in media coverage of environmental issues is unlikely to transform the international political agenda, though it may elicit incremental policy changes. International negotiations are not set up overnight and the complex issues they address cannot be dealt with in the context of a knee-jerk reaction to a peak in the issue-attention cycle. Hence while the media on some occasions act as a catalyst to action, at other times, even simultaneously, they can act as a legitimiser of the agendas and inaction of others. The role of the mass media in the politics of climate change can best be thought of in terms of structuring the debate in a way that actively influences the political response to the problem. It is not so much the setting of agendas as the reinforcement of agendas that makes the media such powerful actors in the politics of global warming. In many ways their influence is secondary, reinforcing the perceptions of primary definers in the politics of global warming: scientists, states and corporations.

5

Climate for business: the political influence of the fossil fuel lobbies

Industry's involvement is a critical factor in the policy deliberations relating to climate change. It is industry that will meet the growing demands of consumers for goods and services. It is industry that develops and disseminates most of the world's technology. It is industry and the private financial community that marshal most of the financial resources that fund the world's economic growth. It is industry that develops, finances and manages most of the investments that enhance and protect the environment. It is industry, therefore, that will be called upon to implement and finance a substantial part of governments' climate change policies.
(International Chamber of Commerce 1995)

5.1 Introduction

In many ways the actions of large industries will determine for decades to come the timing, nature and volume of global greenhouse gas emissions (Schwartz et al. 1992). According to the Business Council for Sustainable Development's figures, 'Industry accounts for more than one third of energy consumed world-wide and uses more energy than any other end-user in industrialised and newly industrialising economies' (Schmidheiny 1992:43).[1] The success of government efforts to forestall the threat of climate change will therefore depend on cooperation with industry. As Levy (1997:56) puts it, 'if an agreement cannot be crafted that gains the consent of major affected industries, there will likely be no agreement at all'. This fact alone makes analysis of the role of industry in climate politics pertinent. Yet despite the significance of corporate actors in the political debate on global warming, emphasised in the above statements, very little academic attention has thus far been paid to the role of business lobbyists in international cooperation on the environment, less still in relation to the issue of global warming specifically.[2]

---

[1] Greenpeace International (1998) has shown in a comparison of CO_2 emissions from the burning of fossil fuels by oil majors with country emissions from fossil fuel combustion, that Shell emits more than Saudi Arabia, Amoco more than Canada, Mobil more than Australia and BP, Exxon and Texaco more than France, Spain and the Netherlands.

[2] Work by Benedick (1991a), Maxwell and Weiner (1993) and Chatterjee and Finger (1994) in other relevant areas, however, has helped to highlight the importance of corporate actors in global environmental politics, even if the significance of their work has not been acknowledged in mainstream regime accounts. Levy and Egan (1998) offer the only other analysis of the influence of the fossil fuel lobbies.

The focus of this chapter is the role of the fossil fuel lobbies in the politics of global warming. It seeks to assess the degree of influence that the lobbies have been able to exert over the course of the international community's climate policy strategy. For the purposes of the analysis here, the fossil fuel lobbies are taken to mean globally organised corporate coalitions representing the coal and oil industries, as well as heavy industry and the car and chemical sectors, which have a key interest in the form that climate policy takes.[3] The fossil fuel lobbies (as opposed to the nuclear energy and renewable energy industries, which are increasingly mobilising a political presence in the debate) are focused upon in order to lend greater diversity to the range of actors covered in this book. They provide the traditional voice of industry in the debate, and are not aligned with the other groups looked at here (such as environmental groups) in the way that the insurance and 'sunrise' (renewable) energy lobbies are. The focus on the fossil fuel lobbies implies particular attention to US lobbies and other Western coalitions of interests. In many Southern states, industrial lobbies have not yet formed around the issue of global warming, though many seem to think that this will change in the coming years (Blantran de Rozari 1996; Narwada 1996; Ratnasiri 1996; Supertran 1996).[4]

In many ways the type of policy under consideration influences the nature of the actors that are drawn into the debate (Greenwood and Ronit 1994). The threat of state action that may result in regulation of the production and consumption of energy, has brought together in loose alliance[5] a range of industries which are involved in the large-scale use and supply of energy that are among the most powerful, wealthy and influential of all sectors in the global political economy (Vaughan and Mickle 1993).[6] The broad consensus among the fossil fuel industries is that action to reduce emission of greenhouse gases, on anything like the scale suggested by the majority of climate scientists, would be highly undesirable given its potential to affect fundamentally the way in which they currently operate. Hence whilst some of the major energy companies have sought to diversify their portfolios and have even invested in renewables, 'The major energy companies still derive the vast majority of their revenues from business lines that would be hurt by emission controls' (Levy 1997:61). A survey of the literature published by the lobbies reveals an

---

[3] For an overview of the perspective of industries centrally involved in the climate policy debate see Vaughan and Mickle (1993), SMMT (1990, 1992, 1993), International Chamber of Commerce (1992), Eurelectric (1991a; 1991b), Bundesverband der Deutschen Industrie (1990), Confederation of British Industry (1993), International Business (1992), Global Climate Coalition (1995a, 1996) Hall et al. (1990), International Federation of Industrial Energy Consumers (1996), World Coal Institute (1998).

[4] Ratnasiri (1996) notes that the Edison Electric Power Institute and the World Council for Sustainable Development (WCSD) have been in contact with Ministries of Trade and Commerce from the South in an attempt to draw them into the debate on global warming. Interview, 16 July 1996.

[5] Other explanations for the political mobilisation of the lobbies include the costs to energy industries of transition to trading in cleaner technology or diversifying into energy service sectors (Steen 1994), the active lobbying of environmental NGOs (Bennett 1996) and the need to coordinate pressure on a number of national governments (Eberlie 1993).

[6] Electricity supply industries in particular play a 'strategic and powerful role in all countries', and are therefore able to exert 'considerable influence in government circles' (Vaughan and Mickle 1993:72).

agenda characterised by an emphasis on scientific uncertainties (Global Climate Coalition 1995b; Gregory and Harrison 1995), the damage to the economy that mitigative action might inflict (Anderson 1996; *ECOAL 4*),[7] and the need to include all greenhouse gases, as opposed to just carbon dioxide, in policy measures. There is also emphasis on carbon sinks and emissions from Southern hemisphere countries, and the role of technological innovation, as opposed to changed patterns of production and consumption.[8] The ultimate goal of more conservative groups such as the American Petroleum Institute is to ensure that 'climate change becomes a non-issue' (Greenpeace International 1998).

Tackling the issue of global warming, unlike issues such as ozone depletion, is potentially about the reduction of production and consumption (dissipating business) and not the substitution or replacement of an offending substance (different business) (Rowlands 1995:137). It poses far more of a threat to the workings of contemporary capitalism, with 'its ties in history, ideology and infrastructure to cheap fossil fuels' (Athanasiou 1991:9), than perhaps any other environmental issue. As Skjaerseth (1994:37) notes, 'confronting climate change requires profound changes in the energy sector which affect most sectors in a modern economy'. This of itself decreases the likelihood of meaningful international action, according to Conca (1993), who argues that it is generally only those environmental initiatives which do not threaten the interests and routines of industrial capitalism that succeed.

5.2   Structural factors/bargaining assets

The coalitions of fossil fuel interests registered at the climate negotiations are just the visible tip of a 'vast bedrock of industrial power' and influence (*ECO*, issue 1, Chantilly 1991). Industrial interests enjoy a number of comparative advantages over other interested parties in the discussions on climate change. Firstly, the operating scale of the lobbies means that they are able to organise themselves to influence policy wherever their interests are threatened (Grant 1993a; Greenwood et al. 1992).[9] Secondly, the comparative weakness of industry groups that dissent from the view that mitigative action on global warming would be damaging to their

---

[7] Richard Briggs, vice-chairman of the Global Climate Coalition (see below), warned Congress in a testimony that abatement measures could impose massive costs on the US economy in the region of $95 billion a year (Dunne 1992).

[8] See for example the joint statement of international business delivered at the 5th Session of the Intergovernmental Negotiating Committee in New York; February 1992, *ECOAL 2*, *ECOAL 4* and Davis (1989).

[9] An example of this would be the EU Committee of the US Chamber of Commerce lobbying against the EU carbon tax (*Environment Digest* 1994). Lobbyists can sometimes exert more influence in policy centres further afield. McLaughlin and Jordan (1993:146) and Grant (1993b) show for example how US groups are among the most effective lobbyists in Brussels.

interests, means that the fossil fuel lobbies have a near monopoly over the industry perspective.[10] Thirdly, industry groups benefit from the fact that international organisations such as the International Chamber of Commerce have an established global presence, a great deal of experience in coordinating their efforts at the international level and a range of contacts with key figures in international institutions.

Fourthly, the structure of the consultation process, set up in the wake of the United Nations Conference on Environment and Development (UNCED), benefits groups that are able to speak with one voice from a common agenda (Chatterjee and Finger 1994:113), such as the fossil fuel lobbies, which have clearly defined interests in relation to climate change. Hence although there are clearly different strategic interests among the fossil fuel lobbies, many companies in this sector have a history of cooperation (in refining and marketing operations), and can subscribe to a common agenda about which sort of political action is desirable and which is threatening to their core interests.

Fifthly, industry lobbyists were permitted special access to the secretariat of UNCED.[11] At the national level too, close contacts often exist with senior figures in influential bureaucratic sectors in government, such as Departments of Trade and Industry, Energy and the Treasury (Wilson 1990):[12] if not ministers themselves, then those reporting directly to them (Vaughan and Mickle 1993). Well-resourced business organisations have channels of access to the very highest echelons of decision-making (Finer 1971; Grant 1993a).[13] There is often a 'rotating door' between the business and policy worlds. The UK government task force on climate change is headed by Lord Marshall, former president of the CBI, an organisation lobbying against the government's energy tax plans (Cowe 1998). Lord Simon of BP was also appointed as a UK government minister for trade and competitiveness in Europe.

Sixthly, the fact that states depend on the activities of industries to a greater extent than other actors in the debate, provides industries with a further source of influence (Newell and Paterson 1998). Influence is therefore exerted not only directly through formal channels of representation, but also structurally, 'because of the crucial role boards and managers exercise over the production, investment and employment decisions which shape the economic and political environment in which governments make policy' (Marsh and Locksley, quoted in Grant 1993c:32). As

---

[10] There is potential for this to change as insurance companies become more involved in the climate negotiations in order to lobby for more action on the issue (Baird 1996; Brown 1996a; Hertsgaard 1996; Leggett 1995a, 1995b). The heavy presence of the nuclear power lobby at the COP4 meeting was also notable.

[11] It is notable that the head of UNCED, Maurice Strong, appointed the chair of the Business Council for Sustainable Development, Stephen Schmidheiny, as his personal adviser (Chatterjee and Finger 1994; Finger 1993; Hildyard 1993).

[12] Boehmer-Christiansen and Skea (1991) demonstrate the close relationship between energy interests and government departments on the acid rain issue, while Hamer (1987:21) describes 'the blurred line that barely exists between the road lobby and certain sectors of the department of Transport'.

[13] For example the steering committee of the European Business Round Table meets twice a year with the president of the Commission and five or six other commissioners to discuss current policy problems (Grant 1993a).

Robert Reinstein (former head of the US delegation to the climate negotiations and industry lobbyist) comments, 'When GCC [the Global Climate Coalition], which represents companies constituting a very significant proportion of the country's GDP starts making noises, they obviously get attention' (Reinstein 1996). The uniqueness of industry lobbyists in this important respect provides the basis for their claim that they should have a separate, UN-recognised consultative mechanism with governments to provide advice on the economic impact of proposed policy strategies (Business and Industry 1996; ICC 1996). It was notable at COP2 (the Second Conference of the Parties to the Convention) that many governments were sympathetic to the idea of providing industry with an extra, exclusive channel of access for this purpose.[14] There is particular sensitivity to the energy sector because 'national governments recognise their structural dependence on the economic health of key sectors' (Levy 1997:63). Suppliers of energy are protected for a variety of security and economic reasons. The association between increased energy consumption and economic growth has a long history (Newell and Paterson 1998). Protection of these industries takes a number of forms, one of which is the use of subsidies. A Greenpeace International study (1998) found that the US government provides a variety of tax breaks and subsidies to the oil industry, totalling $11.9 billion in 1995, and that oil companies pay substantially less than the standard rate of corporate tax.

Seventhly, it is easier to avert international action than to obtain consensus for it, a factor that benefits those seeking to delay orchestrated action. In other words, stalling the machinery of international diplomacy is easier than attempting to craft an agreement that satisfies the interests and addresses the objections of a wide range of conflicting actors and agendas (Porter and Brown 1991:65).

Finally, the financial resources that the lobbies have at their disposal (Brown 1996c; Greenpeace International 1998) enable them to press their case more effectively as they are able to employ the best lobbyists and operate in a more professional manner. This is especially important in countries with a history of interest group involvement such as the US. Such resources have been used by companies to fund both PR firms to lobby on their behalf and 'front' groups to encourage public scepticism about the threat posed by climate change.[15] To bring into sharp relief the difference between the financial resources of environmental groups and industrial groups respectively, the American Petroleum Institute paid the PR firm Burston-Marsteller $1.8 million for a 'grass-roots' letter and phone-in campaign against a proposed tax on fossil fuels. This is more than the combined amount spent by the Environmental Defence Fund, the Sierra Club Natural Resources Defence Fund, the Union of Concerned Scientists and the World Wildlife Fund on their climate

---

[14] The government of New Zealand has been at the forefront of this initiative.

[15] Groups such as The Western Fuels Association, The Edison Electric Institute and the National Coal Association have created a public relations group called the Information Council for the Environment, which launched a $500,000 advertising campaign to 'reposition global warming as theory (not fact)' (*Gallon Environment Letter* 2 (16) 1998).

campaigns (*Gallon Environment Letter* **2** (16) 1998). The scale of the financial resources that the lobbies have been able to bring to bear to press their case has enabled them to support the work of climate sceptics in the scientific community such as Robert Balling, Fred Singer and Patrick Michaels, as well as to finance the development of economic models that predict enormous costs in the event of 'premature' political action (Ozone Action 1997).[16] Contributions to party funds open another channel of access. Fossil fuel contributions to President Clinton's campaign totalled $130,000 in the 1995–6 cycle and Political Action Committee contributions to members of the Senate Energy and Natural Resource Committee totalled approximately $200,000 (ibid.).[17] All these factors contrive to cast a powerful role for industrial groups in global climate politics.

Section 5.3 below focuses on the empirically observable influence that the fossil fuel lobbies exert on international climate policy. The subsequent Section (5.4) is not strictly confined to the visible demonstrations of impact, but also draws on the literature on non-decision-making scenarios (outlined in Chapter 2) to suggest indirect and tacit types of influence. The analysis in Section 5.3 refers to notable changes of policy resulting from lobbying and pressure by industry groups. Section 5.4 discusses both the commonality of interests and mutual dependencies between states and the lobbies, and anticipation by governments of the reactions by industry to particular proposals, which condition the scope of policy. These influences are perhaps harder to identify, but no less significant; a point reinforced by the fact that the industrial lobbyists themselves emphasise the difficulty of tracing the direct effects of their activity and the importance of influencing that is 'done quietly'.[18]

5.3   Observable and direct influence

To assess the nature and extent of the influence of the fossil fuel lobbies over the emerging regime to confront climate change, and to draw out conclusions about stages in the regime's history when influence has been heightened or reduced, a three-stage breakdown of the regime process will be employed, consistent with other chapters in the book: agenda-setting, negotiation-bargaining and implementation.

---

[16] Patrick Balling has received over $200,000 from coal and oil interests for his climate research (Ozone Action 1997). Similarly an economic model developed by Charles River Associates and funded by the American Petroleum Institute was referred to widely in the media and policy discourse within the US (ibid.).

[17] Greenpeace International (1998) estimates that between 1991 and 1996 the oil and gas sector donated $53.4 million to US election candidates and their political parties

[18] Interviews with Walcott, Mulligan and Reinstein, 18 July 1996.

Agenda-setting

Agenda-setting refers to the initial process of policy initiation in response to a perceived problem. It is at this stage that the scope of appropriate action is decided upon, often at the national level in the first instance. Governments seek the opinions and expertise of interested parties at the very start of the policy process, when tentative proposals are being formulated.[19] In a formal sense, this takes place through participation in policy committees.[20] Santaholma (1995:7) notes that 'consultations with business have been an integral part of most national policy processes' in relation to climate change.

Privileged access by lobbyists to government in the earliest stages of policy development can be critical in determining the shape of the policy that emerges. Some of the organisations involved in lobbying on global warming, such as the Union of Industrial Employers' Confederations (UNICE) in Europe, are 'invariably consulted by the Commission almost as a matter of right' (Collie 1993:225).[21] It is often easier for lobbyists to influence climate policy through pressure on their national government, given that they are dealing with just one government, with which they are more likely to have an established relationship (Porter and Brown 1996). National policy processes often offer more scope for participation. As John Shlaes of the Global Climate Coalition (GCC) notes, 'In the US we have dozens of hearings on each issue, here at the UN you don't have those opportunities for representation in the discussion'.[22] Lobbyists also recognise that influencing the scope of policy at the national level is an effective way of shaping the international agenda[23] (Santaholma 1995). As Christophe Bourillon (formerly) of the World Coal Institute (WCI) notes, 'most of the work has to be done on a national level before the negotiations, because governments go to New York or Geneva with a brief'.[24]

Negotiating positions begin to be formulated prior to the negotiation process, and it is here – during the earliest stages of policy development, when the scope of action is considered – that industry groups are able to exert a great deal of influence. The US provides an example. By 1989 the US delegation had already been instructed about their objectives for the UNCED process: a weak framework convention, free

---

[19] See for example *ENDS* (1992d) and Porter and Brown (1996) for more on the way in which the UK Department of Trade and Industry actively encouraged industry to express its concerns about the proposed European carbon tax. A more recent example is the June 1995 European Conference of Ministers of Transport, which held close consultations with the international automobile manufacturing industry in order to produce a joint declaration on the reduction of CO_2 emissions from vehicles (Morand-Francis 1995; *Acid News* 1995).

[20] Questionnaires from Anderson (1996) and Bennett (1996).

[21] Collie (1993:218) shows how UNICE can 'virtually guarantee access on any issues of relevance to European industry'.

[22] Interview with John Shlaes, 18 July 1996.

[23] Interview with Bourillon, 22 June 1995.

[24] Ibid.

of proposals that would reduce the use of fossil fuels (Hatch 1993:13). This is borne out by a leaked 'Talking Points' memo circulated to US negotiators outlining which issues to focus attention upon and which to deflect attention away from.[25] The wording of the memo suggests that the negotiating space available to the US had been predetermined by a concern not to damage the interests of the fossil fuel industries. It encouraged delegates to focus on the many remaining scientific uncertainties and ambiguities in relation to cost; a position that, as Hildyard (1993:29) points out, 'faithfully reflected the position of the oil industry'. Evidence of this influence led NGO observers to note that, while industry groups do engage in formal lobbying during negotiations, 'for the most part they can get what they want in private, months before the bell rings to call delegates into the negotiating sessions' (*ECO*, issue 5, Geneva 1991).

Influence over the scope of the agenda depends in part upon the ability of lobbyists to locate sympathetic individuals in government administrations who will articulate their interests in policy debates.[26] Access for the coal and other lobbies to the US administration was made easier, for example, by the presence of White House Chief of Staff and greenhouse sceptic John Sununu, who was in a position to decide which views on the costs of abatement action President Bush should be exposed to.[27] Acting on behalf of powerful departments within the administration – such as the Office of Management and Budget (OMB) and Council of Economic Advisors (CEA) – Sununu was able to gut legislation that threatened fossil fuel interests.[28] Hence when the Department of Energy revealed its national energy strategy in February 1991, as a key part of the US administration's climate change policy, a number of major proposals, including increased corporate average fuel efficiency standards (CAFE) and energy conservation, had been removed (Hatch 1993). Instead the strategy included a proposal to allow increased oil and gas production in Alaska (Eikeland 1993b:989). There was little mention in the strategy of coal and oil as such, apart from research and development provisions on clean coal technology (which the Clean Coal Technology Coalition had lobbied for and which passed through Congress without debate or change), and tax relief for independent petroleum producers (Eikeland 1993b:992; Hatch 1993). All provisions directly relating to action on climate change met a veto threat and were 'extensively modified' in the final bill (Eikeland 1993b:993). Hatch (1993:19) refers to the 'intense efforts of industry', particularly the coal, oil and car industries, when explaining the assault on the strategy. A broad alliance consisting of the National Coal Association

---

[25] The 'Talking Points' memo (dated 17 April 1990) is reproduced in *ECO*, issue 7, Geneva 1990.

[26] Interview with Mulligan, 18 July 1996.

[27] Similarly, John Dingell, chairman of the Energy and Commerce Committee and described as one of the 'most powerful figures in congress' mobilised 'unyielding opposition' to the passage of any energy bill that threatened the interests he represented (Hatch 1993:21). Richard Darman, director of the Office of Management and Budget, was also a key opponent of any international commitments on global warming (ibid.)

[28] For more on Sununu's role in vetoing proposals for greater action on global warming on behalf of the Office of Management and Budget, the Council of Economic Advisers and the Treasury, particularly in relation to the 1992 Congress Energy Policy Act, see Eikeland (1993a).

(NCA), the GCC[29] and the Climate Council successfully resisted all proposals designed to stabilise or reduce CO_2 emissions (ibid.). An indication of the closeness of the relationship that the lobbies were able to nurture with the White House chief of staff was the *ECO* headline announcing Sununu's resignation: 'Sununu resigns . . . Coal lobby in mourning' (*ECO*, issue 1, Geneva 1991).

The lobbies also had a sympathetic ally in President Bush, who made it clear he would veto any bill that sought to impose new taxes or bar new oil and gas production (Eikeland 1993a:5; Hatch 1993). Nevertheless the case of the US illustrates how the lobbies have to try to influence governments in different ways,[30] and that they enjoy better relations with some government administrations than others. Hence the influence of the lobbies in the US may be said to have has declined slightly under Clinton's leadership compared with the degree of access they enjoyed under the Bush administration.[31]

The EU proposal for a community-wide carbon tax perhaps provides the clearest example of the strength of the influence of fossil fuel lobbies in vetoing climate policy instruments that threaten their interests.[32] The prospect of the tax 'spurred the massed ranks of Europe's industrialists to mount what is probably their most powerful offensive against an EC proposal' (*Economist* 1992a; *ENDS Report* 1992c).[33] Carlo Ripa di Meana, EU Environment Commissioner at the time, described the lobbying as a 'violent assault'; an indication of the vigour with which the industries pursued their demands (quoted in Sebenius 1994:294). It is clear that the arguments forwarded by industry featured highly in policy-makers' minds during the formulation of the tax proposal. Philip Dykins of the Department of the Environments' Global Atmosphere Division and representative of the UK government at the climate negotiations, stated that 'we're very concerned about the possible impacts of a carbon tax on industry . . . it's a competitive disadvantage for industry, that's a big consideration'.[34] Lobbying by the UK coal industry against the tax was successful in bringing on side the House of Lords EU Committee, which produced a critical report on the effect the tax would have on the competitiveness of the coal industry (*Environment Business* 1992a). The committee was headed by Lord Ezra, a former chairman of British Coal (ibid.).

---

[29] The GCC describes itself as 'the leading business voice on climate change'. It is a organisation of over fifty-five business trade associations and companies, including the American Petroleum Institute, the US Chamber of Commerce, Du Pont, Dow, Exxon, Texaco, Chevron, Mobil and a number of car (General Motors, Chrysler) and road construction companies. The coalition was established in 1989. Brown (1996a) describes the GCC as 'the giant of the lobbying business'.

[30] Interviews with Reinstein and Walcott, 18 July 1996.

[31] Interview with Reinstein, 18 July 1996.

[32] For a useful collection of position papers from the leading industrial actors on the carbon tax issue see Forum Europe (1992), Hampson (1992) and UNICE (1994).

[33] As *The Economist* (1992a) also wrote at the time, 'The proposed carbon tax has been subject to the most ferocious lobbying ever seen in Brussels'.

[34] Interview with Dykins, 25 February 1994.

Despite the fact that EC commissioners were told that in order for the community to meet its goal of stabilising emissions at 1990 levels by the end of the decade, a US$18 a barrel tax would be necessary, they concluded that the tax was 'not politically realistic' (quoted in *ECO*, issue 2, Nairobi 1991). The development of the EU carbon tax into a conditional proposal,[35] dependent for its approval on similar actions being adopted by industrial competitors in the US and Japan, and from which energy-intensive industries were in any case exempt, provides clear evidence of the impact of the fossil fuel lobbies on the course of international climate politics. Skjaerseth (1994:31) argues that 'inclusion of the principle of conditionality must be seen in the light of increasing lobby activity from business interests'. For some observers, these significant caveats to the proposal revealed a desire on the part of governments to 'limit damage to oil companies which argue the tax would handicap them in international markets' (Gribben 1992). The Commission responded to the concern of the oil industry, vocalised for the most part through the European Petroleum Industry Association (EUROPIA), by demanding that officials prove that the tax would be the most economically efficient way of meeting the stabilisation target (*Environment Business* 1991). The tax proposal ended up not even as a draft directive, but as a recommendation for legislation, with each EU member deciding individually whether or not to implement it (*ECOAL 15* 1995; *Environment Business* 1991b:5). Lobbyists were able to erode the consensus among EU member states in favour of a binding directive (*Guardian* 1992a; Palmer 1992:92). There is no doubt about the cause: 'The commission significantly watered down Mr Di Meana's original proposals following heavy pressure from industry groups throughout the EC' (*Guardian* 1992a; *ENDS Report* 1992d). Indeed 'agreement on the proposal only became possible after fervent lobbying by European industry groups produced a much watered down version of the energy tax than the commission originally wanted' (*Environment Business* 1991b).

Hence it was not just a question of the lobbies influencing the course of the proposal once laid out, it was a question of pre-empting its contents and therefore helping to set the agenda. Sebenius notes that, 'Largely as a result of industry opposition, before the carbon tax was even proposed as a directive to the council of ministers, both energy-intensive industries and major exporters were *pre-emptively* exempted from the tax' (Sebenius 1994:294, emphasis added). The fate of the proposal to tax carbon vividly demonstrates the ability of the fossil fuel lobbies to set the boundaries for discussion of climate change strategies.

Industry groups can also pre-empt, and render untenable, particular negotiating positions by threatening to relocate their business if a government pursues a particular line. The Dutch energy industry's opposition to the Dutch government's proposal to introduce a unilateral energy levy, took the form of a threat to transfer new investments abroad if the government developed its plans further (*ENDS Report* 1992c; *Environment Business* 1992d). These industries also refused to participate in

---

[35] Australia and Japan have also adopted an emission reduction target that is conditional on their major trading partners adopting a similar target (Morand-Francis 1995).

environmental planning discussions with the government unless the latter dropped its proposal to double energy taxes (*Environment Business* 1991e).[36] The government subsequently postponed its decision on the levy until completion of the distant EU-wide proposal (*Environment Business* 1992b). Hence the proposal was effectively terminated at birth by protests from industry (*Environment Business* 1991b, 1992d, 1992e).

The Byrd Resolution (Senate Resolution 98), sponsored by Republican senators on behalf of the GCC, effectively tied the hands of the US negotiators at Kyoto by declaring that the (Republican-dominated) Senate would not ratify any agreement emerging from the conference that did not explicitly include reduction commitments for LDCs. The tactic was successful in restricting the degree of negotiating space available to the US delegation at the Kyoto meeting in agreeing new targets.

Internalisation of the concerns of the fossil fuel lobbies by leading government officials will therefore delineate the possibilities for future action at the international level. Governments take with them to the UN negotiating chambers a sense of what agreements will be acceptable to influential domestic constituencies, and which proposals are likely to provoke the sort of furore already witnessed at the national level.

Negotiation-bargaining

The direct influence of the fossil fuel lobbies upon states once international negotiations have commenced may be thought to be minimal, given that the scope for negotiating compromises and concessions is to some degree determined beforehand, leading some observers to claim that the power of industry lobbies during international negotiations is slight.[37] The underlying assumption is that once policy has shifted to the negotiating halls of the UN, further resolution of the issue is solely the domain of those government officials and diplomats whose job it is to draft appropriate texts. The types of influence that business groups are able to exert at the national level are far less effective at the international level, given that international organisations are more insulated from the pressure lobby groups can bring to bear in the domestic context (Levy and Egan 1998).[38] There were, moreover, few opportunities for industry representatives to speak on the floor of INC plenary sessions (Faulkner 1994).

However, the new openness during the UNCED negotiations meant that representatives from industry were included as full members of several national delegations (Chatterjee and Finger 1994; Faulkner 1994). In the case of the US, it was not just a question of industry representatives being conferred access to government.

---

[36] Among those making the threat were oil and chemical companies such as Shell, Akzo, DSM and Dow.

[37] Interview with Robinson, 15 July 1994. As the environmental NGO newsletter *ECO* claimed, 'The professional lobbyists of the coal, oil and car industries are here of course in Geneva. But in truth their role, like that of the environmental groups, is now minimal' (*ECO*, issue 1, Geneva 1990).

[38] Though channels of influence such as social networks, media advertising and campaign contributions are not available at the international level, spillover effects should not be underestimated.

Highly placed officials from the Bush administration also appeared on industrial lob-bying delegations at INC meetings (Nilsson and Pitt 1994:141). Indeed Robert Reinstein, former head of the US delegation to the climate negotiations under President Bush, has attended recent rounds of climate negotiations as a representa-tive of the Canadian Electricity Association. John Schlaes, director of the GCC, also held a senior position in the executive office of the White House as director of communications under John Sununu (Levy and Egan 1998), and on leaving her post as senior director for Global Environmental Affairs at the National Security Council, Eileen Clausen took up a post with the International Climate Change Partnership (ICCP). Hence the 'revolving door' between the corporate and govern-ment worlds is also manifested at the international level.

Moreover the drafting process and the need to develop compromises acceptable to all parties, creates fresh opportunities for policy shifts and changes in direc-tion. Industrial lobbyists attend negotiations in order to influence these changes. As former WCI lobbyist Christophe Bourillon says, 'We are there for the last minute drafting and changes. When states invite you to help them draft something you don't say no'.[39] William Mulligan of the GCC says of the US delegation, 'they will turn to us to find out what kind of implications there are for the private sector of any new initiative'.[40] This reinforces the point that industry groups are regarded as being able to provide governments with a sense of whether particular policy proposals are realistic or feasible in an economic sense.[41] This process of referral allows the lobbyists further scope to determine the range of options under conside-ration. In the words of Tito Sale of the World Energy Council, 'Our role is to advise governments about *their* industrial and technological flexibilities.'[42] The leverage provided by these forms of consultations helps to explain why, for Robert Anderson of the American Petroleum Institute, contact with diplomats at interna-tional negotiations is the most effective channel of influence available to corporate lobbyists.[43]

One other means of exerting influence at this stage of the policy process, which lobbyists are open about using, is to remind delegations that whatever they agree in the negotiating halls will have to pass through national legislatures that may be hostile to the proposals.[44] The GCC and Climate Council for example, sought to restrain further actions, circulating letters expressing concern by members of the US Congress and Senate about the course being pursued by the US delegation at the COP2 negotiations. Press conferences were also held at Kyoto and subsequently at COP4 in Buenos Aires to reiterate the veto threat implied by the Byrd Resolution (see above) in order to maintain pressure on the US delegation not to

---

[39] Interview with Bourillon, 22 June 1995.
[40] Interview with Mulligan, 18 July 1996.
[41] Interview with Walcott, 18 July 1996.
[42] Interview with Sale, 18 July 1996, emphasis added.
[43] Questionnaire from Anderson (1996).
[44] Interviews with Reinstein and Schlaes, 18 July 1996.

accept accords without commitments from non-Annex 1 parties. In this way the lobbies are able to undermine the prospects of policy options that they consider to be too costly or unrealistic, by bringing domestic political pressures to bear in international arenas.

Lobbyists also adopt new strategies to influence the negotiations: grouping together[45] and forming alliances with governments. One channel of influence for lobbies during the negotiations has been the formation of coalitions with what might be termed 'laggard' or 'veto' states (Porter and Brown 1991): those states whose principal goal in the negotiations is to thwart action on issues they consider to be contrary to their interests. The fossil fuel lobbies have been able to advance their minimal action agenda during the negotiations, through support for the negotiating position of oil exporting states such as Saudi Arabia and Kuwait in the Organisation of Petroleum Exporting Countries (OPEC) bloc (*Acid News* 1994b; Porter and Brown 1996). Don Pearlman of the Climate Council and John Schlaes of the GCC were reported to have drafted a number of US–Saudi amendments designed to stall negotiations on a protocol to the convention (*ECO*, issue 8, New York 1993). For instance the proposal by Saudi Arabia that protocols to the convention should be adopted by three-quarters of the parties instead of the present two-thirds was 'widely believed to have been drafted by US fossil fuel lobby interests' (*ECO*, issue 3, New York 1994). This delaying tactic was said to have had 'Pearlman's fingerprints all over it' (Farley 1997). Sathiah Renji, head of the Malaysian delegation at the climate talks, said of Pearlman, 'He has tremendous influence and countries depend on him. I've seen fossil fuel producers consult with him before making a decision' (Farley 1997). Renji recounts an incident in which a Middle Eastern delegate asked for a vote at a closed meeting to be delayed while he went to the bathroom. The chair sent an escort along and found Pearlman waiting outside (ibid.). Indeed it was the lobbying antics of Don Pearlman at INC 11 that resulted in NGOs being banned from the floor during the negotiating sessions. Pearlman 'was standing there writing out interventions on pieces of paper that his runners were then taking to his client states in such a blatant way that it led to all the NGOs being banned'.[46] The ties run both ways however, and Grubb et al. (1999) argue that much of the negotiating strength of the OPEC group in the climate negotiations derives from their links with US-based industries.

Alignment with 'veto state' partners ensured the deletion of key sections of draft proposals of what eventually became the Climate Convention. For example reference to the Toronto target of a 20 percent reduction in CO_2 emissions by the

---

[45] The need to group together to have more impact at the international level provided the impetus for the creation of the World Industry Council for the Environment (WICE) after UNCED in 1992, an amalgamation of the lobbying efforts of the ICC and BCSD. The BCSD and WICE were merged early in 1995 to form the World Business Council for Sustainable Development, which is a coalition of 120 companies drawn from 35 countries with a network of national and regional business councils.

[46] Interview with Sieghart, 17 July 1996. The Climate Council, which Pearlman heads, is said to have greater influence with the OPEC states, while the GCC has closer relations with the JUSCANZ grouping (Levy and Egan 1998).

year 2005 in a draft declaration at the Second World Climate Conference in 1990 was removed after lobbying by the US, the former Soviet Union and Saudi Arabia, purportedly on behalf of their oil lobbies (*ECO*, issue 7, Geneva 1990). As Alden Meyer noted at the time, 'The US . . . is working with the fossil fuel industries to block any meaningful action by the ministers' (*ECO*, issue 7, Geneva 1990). For Doyle (1992a), the fact that the convention contains no binding commitments to act, illustrates that 'the industry lobby won the day'. Rumours during the negotiations that the US administration might drop its resistance to mandated reductions were dispelled after coal and other lobbyists 'intervened forcefully' to check a policy shift in that direction (Dawson 1992).[47] Governments' concerns for the energy industries explicitly worked their way into the text of the convention. Article 4 (part 10) on commitments provides an exemption to countries whose economies are 'highly dependent' on producing or consuming fossil fuels (UNFCCC 1992:4(10), 4(8)). The WCI and other groups were instrumental in the drafting of this proposal, in collaboration with Australia and the OPEC group.[48]

At the First Conference of the Parties to the Climate Convention (COP1) in Berlin, the International Climate Change Partnership (ICCP),[49] a group of corporate lobbyists representing a number of fossil fuel interests, provided the US delegation with a marked-up version of a draft document that formed the basis of the final conference document (Fay and Stirpe 1995). Many of the concerns articulated by the ICCP were replicated virtually verbatim in the final conference text. Even the ICCP's preferred language was adopted. The ICCP's preference for reference only to 'a legal instrument', so as not to privilege discussion of a protocol, was included in the preamble of the final conference text.

The lobby was also successful in removing from paragraph 2(e) of a draft conference text mention of 'common measures', so as to privilege domestic policy options over which the ICCP would be able to exert more influence. Perhaps more interesting still is the way in which the US Senate's Committee on Energy and Natural Resources responded to the outcome of COP1. The committee criticised the conference in terms that resonated exactly with the criticisms levelled by the GCC (1995a).[50] This is perhaps less surprising when it is recalled that 60 percent of the

---

[47] According to an *ECO* editorial, 'The Climate Council and the Global Climate Coalition have opposed any commitment even to limit emissions, arguing for a purely procedural convention . . . they have largely gotten their way' (*ECO*, issue 8, New York 1992).

[48] Interview with Bourillon, 22 June 1995.

[49] The ICCP, established in 1991, describes itself as a 'diverse industry organisation dedicated to responsible participation in the international and domestic global climate change policy process' (ICCP brochure undated). Members include DuPont, General Electric, BP and Enron. The group claims to represent the 'middle ground' in the debate (Levy 1997).

[50] See the press release of the Senate Committee on Energy and Natural Resources (1995) and The Global Climate Coalition's press briefing (Shlaes 1995). A similar pattern of events emerged at COP2, where the US Senate Committee on Energy and Natural Resources published a press release and the US Senate and Congress sent letters to the US President and Secretary of State, which drew on statistics and arguments identical to those contained in the Global Climate Coalition's press briefings released shortly before this (Murkoweski 1996; US Senate 1996).

committee were from coal states, 'with oil and gas amply represented as well' (Hatch 1993:20). ICCP executive-director Kevin Fay, was able to conclude from COP1 that, 'The parties met the criteria for success from our perspective' (ICCP 1995a).

It was also apparent at COP1 that what Grubb (1995a:81) refers to as a 'hardcore' of states, resisting further action and insisting on commitments for developing countries, was emerging in the form of the US and Australia, which perhaps uncoincidentally also had 'the biggest and most powerful domestic fossil fuel (particularly coal) industries' (ibid.). The fossil fuel lobbies, dominated by the concerns of the US coal and oil industries, 'rushed to the negotiating floor' at the final preparatory meeting for COP1 (February 1995) when it looked as though agreement would be reached on the inadequacy of existing commitments (ibid.:182). The result was that 'key exporter governments renewed their objections apparently at the behest of industry lobbyists' (ibid.).

Even in the run-up to the Kyoto agreement, which ended up going against the interests of the fossil lobbies in many respects, 'US industry and in particular the electricity sector ... lobbied strenuously against early commitments. Desperate to mollify at least some of the domestic opposition, the administration let it be understood that it would not accept any emissions reductions binding before 2010 (Grubb et al. 1999). On the issue of which gases to include in the Kyoto Protocol, Greenpeace claimed that 'an agreement between the Dutch government and its powerful chemical industry was at the root of EU opposition to including the trace industrial gases (HFCs, PFCs and SF_6) at all' (cited in Grubb et al. 1999).

Whilst it is difficult to ascertain how far states are pushed by the lobbies, or the extent to which lobbying by industry groups merely serves to support conservative negotiating positions, it is clear that the lobbies push as well as support sympathetic negotiating positions. It is plausible to argue, for example, that the US stance was easier to sustain with the active support of such powerful domestic constituencies, convincing the negotiating delegation that the line they were taking would be popular and tenable 'back home', and at the same time helping to undermine the position of advocates of action. More than that, John Gummer, former secretary of state for the environment in the UK, argues that the fossil fuel lobbyists and the sceptics in the scientific community whose work they fund, 'make respectable some of the wishes of the politicians to avoid difficult issues' (quoted in Ghazi 1997).

The PR firms representing the companies continually seek out for new allies, and when it became clear that Australia would be a willing veto state in the run-up to the Kyoto meeting, major US PR firms (such as Edelmans, Burston-Marsteller, and Hill and Knowlton) increased their presence in Australia by teaming up with the mining industry (*Gallon Environment Letter* 2(16) 1998). According to the *Guardian*, 'Perhaps nowhere has the fossil fuel industry been more successful than in Australia, where the government has presented industry lobby interests as synonymous with the national interest' (Beder, Brown and Vidal 1997:4). The Australian Bureau of Agriculture and Resource Economics (ABARE), a government-funded economic forecasting agency that liases closely with Australia's Western Mining Corporation, the US Mining Association, Ford Motors and the American

Petroleum Institute, predicted huge costs in jobs and income if the emission reduction targets were met (ibid.). The ABARE-sponsored model was used to support the Australian government's negotiating position that the stabilisation goal of the convention would only be achievable if non-OECD parties accepted commitments. Corporations willing to pay $50,000 could buy a seat on the ABARE steering committee (ibid.). In return, as ABARE's promotional publicity for sponsors states, 'By becoming a member of the consortium, you will have an influence on the direction of the model development' (Burton and Rampton 1998). As a result ABARE receives $500,000 a year (80 percent of its overall costs) from companies such as Mobil Oil, Exxon, Texaco and Statoil (ibid.). Hence when US industry lobbies hook up with sympathetic Australian government departments, a strategic alliance is formed that can simultaneously bolster the government's negotiating position and extend the lobbies' influence in new directions.

The lobbies have also been able to create conflict between countries within international fora by playing them off against each other at the national level. A Charles Rivers Associates (CRA) economic forecasting model, funded by the American Petroleum Institute, has been used to support the argument of the US administration that action should not be taken unless LDCs also take action, and at the same time to persuade LDCs that CO_2 reduction strategies will hinder their development (Ozone Action 1997). This tactic helped to achieve stalemate that threatened (and may still threaten) the (long-term) outcome of the Kyoto agreement. Whilst pressing home the need for LDCs to commit themselves to reducing their CO_2 emissions, within the US, as part of the GCC, Exxon Chairman Lee Raymond was at the same time, urging Chinese industry and government officials not to accept emission limits.[51] This 'double-edged diplomacy' was used by Eivind Retien, director of Norsk Hydro in Norway, who spoke out against CO_2 taxes in Norway on the ground that other countries did not levy such taxes, while a fellow director of the company was opposing the adoption of the carbon tax, which would apply such a tax to other EU member states (*Norway Daily*, 24 June 1996).

Attempting to slow the work of the IPCC is another tactic used by the lobbies throughout the negotiation process. The IPCC has become a target because of the sway the lobbies feel it has over politicians.[52] In January 1992, when the IPCC adopted its supplementary report, it developed updated and alternative reference scenarios in a move that was 'widely seen as a concession to the coal industry' (*ECOAL 2* undated). This illustrates that lobbying these bodies that generate the knowledge base of policy enables lobbies to shape scientific input into the debate. The World Coal Institute, the Climate Council, the World Energy Council and other individual members of these and other fossil fuel lobbies, such as Mobil Oil,

---

[51] Raymond said in his address to the World Petroleum Congress in Beijing, 'I hope that the governments of this region will work with us to resist policies that could strangle economic growth' (Greenpeace International 1998:46).

[52] Pearce (1995d) also shows how a GCC-funded forecasting company called Accu-weather published a report at COP1 attempting to discredit the work of the IPCC. Questionnaire from Anderson (1996).

the National Coal Association and Edison Electric Institute, are all reviewers of IPCC Working Group 1 reports (IPCC 1995).

The GCC and Climate Council became embroiled in an open battle with the IPCC over procedure at an IPCC meeting in September 1994 (*ENDS Report* 1994b). The groups were accused by the IPCC of 'nit-picking', of seeking to delay the IPCC's work at every possible stage by raising points of order (after they had sent letters to leading US politicians complaining about abuse of procedure in the IPCC's work), and of forming an unholy alliance with the oil producing states (ibid.). Similarly the Global Commons Institute (GCI) claimed that in the run-up to COP2, scientists in the IPCC were 'yielding to pressure from industry to foresee yet higher atmospheric pollution as acceptable' (Global Commons Institute 1996:1). The GCI's claim is supported by the activities of the World Energy Council, a lobby group representing industries from more than 100 countries, which urged governments not to accept the IPCC recommendations published in June 1996 on the basis that the advice was deficient, unrealistic and influenced by academics seeking to attract funding for their work (*Environment Digest* 1996:9).

Implementation

Commitments made at the international level are of course meaningless unless they are enforced, and some industry groups have made clear to governments their desire for any obligations that are undertaken to be enforced, for fear of a competitive edge being gained by rival companies in a 'free-riding' state.[53] Yet at the implementation stage, the fossil fuel lobbies are also presented with further opportunities to delay the advancement of policies damaging to their interests.

The focus returns once more to the national level, where close relations with government can once again be utilised. In terms of the pace of implementation of greenhouse policies, governments are constrained by the willingness or otherwise of industry to cooperate. Porter and Brown (1996:62) note how the climate regime is 'particularly sensitive to the willingness of corporations in key countries to take actions that would allow the international community to go beyond the existing agreement'. In certain areas, control of energy production by many of the industries that make up the lobbies means that they are, by definition, a force with which governments have to negotiate on issues of energy planning and implementation (Nilsson and Pitt 1994:38).[54] Moreover, in the post-Rio era, in which a dialogue is unfolding between governments and corporations about the most cost-effective way of achieving policy ends, the process of clarifying and implementing the terms of the Climate Convention offers further scope for bargaining and influence over policy. Many of the obligations in the convention are sufficiently ambiguous to enable

---

[53] See ICCP (1994, 1995).

[54] Wilson (1990) shows how in the US in particular, powerful incentives operate to ensure that the agencies responsible for representing particular interests are sensitive to their concerns (such as budget cuts and the prospect of congressional brawls with those interests if they feel aggrieved).

industry groups to press upon governments their preferred interpretations of the commitments. The endorsement of joint implementation (JI) by the Kyoto Protocol and the opportunity to shape the agenda and operation of the newly created clean development mechanism (CDM), open up further avenues of potential industry input and influence. Both JI and the projects promoted and funded by the CDM will require the cooperation and supportive involvement of industry.

Direct lobbying also has a role to play at the implementation stage. Lobbies can get proposals watered down even after governments have agreed to them in international fora. Morand-Francis (1995:67) notes that in the immediate aftermath of the convention, 'it soon became clear to the various countries that implementing the required measures even with the label "no-regret" was no trivial matter'. Sebenius cites the case of Japan, which prior to UNCED agreed to cut its CO_2 emissions to 1990 levels by the year 2000, yet by June 1993 had watered down legislation designed to meet this objective 'in response to strong opposition from Japanese business leaders and the Ministry of International Trade and Industry' (Sebenius 1994:295). Porter and Brown (1996:61) show how, in 1994, the National Association of Manufacturers and the US Chamber of Commerce, in conjunction with the electric power industry, threatened the funding for the US National Climate Change Action Plan in Congress after Clinton supported the negotiation of binding commitments at Berlin (the first meeting of the Conference of the Parties to the Climate Convention) to reduce greenhouse gas emissions beyond the year 2000.

In a similar way, the UK government's plans to raise up to £400 million a year through levies (by the Energy Savings Trust) on gas and electricity bills as part of the government's convention obligations, were successfully resisted by the energy industry and its regulators, Ofgas and Offer (*Environment Digest* 1994:10; Pearce 1995c). O'Riordan and Rowbotham (1996:252) note that 'Established industry . . . pressure groups circulating around DTI and the Treasury could easily see off the tax'. Where governments have sought to put measures in place, influential industries have successfully lobbied for exemption. In Norway the coal tax is limited to coal used for heating purposes, and heavy industries such as the metal and cement sectors are exempted, as are emissions from industrial processing. In Italy the Ministry of Industry and its affiliates, such as the UPI (the Italian Oil Board), said to be a 'powerful lobby in the energy debate', 'created obstacles throughout the process leading to the Italian CO_2 stabilisation plan' and 'succeeded in imposing modifications in the plan itself' (Marchetti 1996:310, 306).

Hence resistance generated as national governments come to implement their international commitments, can stall those proposals that are regarded as threatening to key economic sectors. In the case of the US, Grubb (1995a:89) notes how, despite the presence of environmentalists in the administration, implementation policy is dominated by Congress, where coal-producing states are strongly represented; a factor that militates against the implementation of legislation at the national and state levels. Frank Murkoweski, chair of the US Senate Committee on Energy and Natural Resources, who receives the largest proportion of his corporate sponsorship from the oil and gas sector (Greenpeace International 1997), declared in a press briefing released at COP2 that the 'US negotiating position is likely to result in a treaty amendment or a new protocol that the Senate will be unable to ratify' (Murkoweski 1996). US environmentalist Dan Lashof identifies the 'great influence' of the GCC (in terms of its degree of access to congress and the money it can invest in congressional campaigns) as 'one key reason why Congress is actively hostile and unsupportive to taking action on this issue'.[55] Given this, the Kyoto Protocol faces a tough ride, made more difficult by the Senate's Byrd Resolution, which expressed the unwillingness of the Senate to ratify any agreement that does not contain commitments on the part of less developed countries. The lobbies have an obvious incentive to delay the implementation of the protocol for as long as possible. It has been estimated that even a one-year delay in ratification of the protocol by the US could benefit Exxon's bottom line by $200 million (Boyle 1998).

Deliberate non-cooperation in the implementation stage is a further channel of resistance for the lobbies. A number of energy sectors in Europe warned their governments that they would not proceed to meet voluntary energy efficiency targets if a carbon tax was implemented unilaterally; in effect closing off one of the proposed strategies for meeting the European stabilisation target (*ENDS Report* 1992b). Trade associations of appliance makers also managed to obstruct EU efforts to impose mandatory standards for energy efficiency on their products, by refusing to supply data or cooperate in an energy efficiency study upon which policy would be based, thus delaying the efforts of the EU to meet its obligations under the Climate Convention (*Acid News* 1994e).[56]

The above examples show that proposals can be altered, weakened or overturned altogether. Influence can be exerted in advance of negotiations, as well as when negotiations are under way. Failing that, strategies can be employed to ensure either that the implementation of measures is made as difficult as possible, or that some initiatives are not realised. The political power of the lobbies is apparent at every stage of the policy process on global warming.

5.4   Compatibility of agendas

Business influence is never easy to see behind the scenes, but is tangible in policy outcomes. (Weir 1996)

Using a simple input–output analysis that looks at which of the many agendas win through in the policy debate, it can be observed that the course of action advocated by the fossil fuel lobbies, what may broadly be termed a 'no-regrets' approach, is

---

[55] Interview with Lashof, 18 July 1996.
[56] Proposals for legislation on energy efficiency standards were also fiercely resisted by the powerful European Roundtable of Industrialists, which represents a number of key industries in Europe (see below) (*Environment Business* 1993a).

compatible with the negotiating stance of a number of leading states in the nego-tiations.[57] The approach of the US (one of the most outspoken advocates of 'no-regrets'), with its focus upon net emissions[58] as a means of deflecting attention away from domestic emission reductions, is compatible with the agenda advocated by the lobbies. It is notable, for example, that the GCC was quick to endorse the Bush administration's decision to provide a $75 million fund to assist less developed countries to reduce their greenhouse gas emissions (Global Climate Coalition 1992). Equally in evidence is praise for the US government's 'comprehensive' approach (including all greenhouse gases rather than just CO_2) and its emphasis on voluntary measures (ICCP 1995). This insistence on maximum national flexibility in developing response measures, evident at COP2, was applauded by the US fossil fuel lobbies, which had pressed this point upon the negotiators before and during the COP2 meeting (ICCP 1996). Pressure from companies for flexibility in the run-up to the Kyoto agreement was also apparent in the demand of OECD countries that their companies be able to exploit reduction possibilities internationally (Grubb et al. 1999).

Another key tenet of the post-Rio policy debate has been joint implementation, where higher emitting states can invest in countries where it is cheapest to reduce emissions. Enthusiasm for joint implementation on the part of some Norwegian and US industries has been reflected in their governments' ongoing support for this type of action (Rowlands 1995:159).[59] According to Canadian environmentalist Robert Gendron, 'It is no secret that joint implementation is meant to help some industry dinosaurs survive for a few more decades' (*ECO*, issue 6, New York 1994). Similarly with tradable permits, now firmly on the agenda post-Kyoto, a significant effort has been made by the US government to assure businesses that it will not accept measures that are 'economically disruptive' to them (Levy 1997:65).

Many of the policy proposals emerging from governments have also emphasised voluntary codes, agreements between the private sector and the government (APPEA 1996; Shell Australia 1996) and incremental energy-saving measures; exactly those measures endorsed by the fossil fuel industries. In some cases industries have volunteered emission reductions in return for a government promise to post-pone regulatory measures to combat climate change (APPEA 1996). *Acid News* describes a trade-off where, in response to German industry tightening its voluntary agreements to reduce CO_2, '*For its part*, the [German] government has announced it has no plans to introduce a national CO_2/energy tax and would exempt those parts of the industry that adhere to the voluntary commitment from any EU-wide tax' (1996: emphasis added). A similar situation prevails in the Netherlands: 'We have a lot of voluntary agreements with industry and so we have to liaise closely

---

[57] The 'no-regrets' approach refers to policies that address other concerns, rather than global warming alone. In this way there will be 'no regrets' if the impact of global warming is not as severe as some suggest.

[58] Net emissions refers to total emissions minus total sinks.

[59] Indeed energy interests have initiated their own private joint implementation programmes (Kiernan 1994; Vatikiotis 1994).

with them to see how much potential there is for energy efficiency, and the outcome is what we can agree between us'.[60] Even when governments insist on pursuing measures that energy industries oppose, guarantees may be made that any loss of revenue will be compensated by tax cuts. A UK government proposal for an industrial energy tax was made acceptable to industries by promising that 'any revenue raised must be returned to industry via tax cuts in other areas' (Browne 1998:4; Cowe 1998:16). These examples illustrate a bargaining process taking place between government and industry groups, where the latter are afforded significant leverage in shaping policy options.

Taken on their own, some of the compatibilities between government agendas and those of the lobbies described above, would only tell us so much about the political influence of the fossil fuel lobbies. In combination, however, with the examples outlined above, it is harder to dismiss them as coincidental evidence.

5.5   Cross-issue influence

To assess more fully the scope of the influence of the lobbies, it is necessary to look beyond the formal institutional fora of climate politics. The ability of lobbies to advance their preferences across a range of policy areas, means that the effectiveness of policies on climate change can readily be countered by policies developed elsewhere in government. For instance, the tough 25–30 percent reduction in CO_2 emissions to which the German government has committed itself is unlikely to be achieved given that decisions are being made in energy and transport policy that undermine action on climate change (Bach 1995). The German government's plan to develop a coal mine in North Rhine-Westphalia, known as Gartweiller II, is argued to be 'inconsistent ... with German climate policy declarations' (Grubb 1995a:92). The influence of industry lobbies in areas such as road and energy policy, inhibit the government's room for manoeuvre in respect of climate policy.

Equally, the leadership ambitions of Norway have been scuppered by the plans of the Energy and Industry Ministries to expand production of petroleum and gas, encouraged by the lobbies for those fuels. Sydnes (1996:295) notes that 'Oil and transport policies are still on different tracks, largely unaffected by the climate debate, driven by domestic interests and powerful lobbies.' Whether there will be any change in the UK's negotiating position following the election of the Blair government remains to be seen, given that it is unclear how the government intends to square its CO_2 reduction commitment with a move to save the coal industry and extend the moratorium on gas-fired plants (Boyle 1998). O'Riordan and Jordan (1996:77) note that stable policy communities in the transport and electricity generating sectors are an 'effective constraint to radical policy change' in that governments have to gain the support of these communities, which, as many of the cases above illustrate, has not been forthcoming.

[60] See Cozijnsen (1996).

The US also had its hands tied over the proposal for an international protocol on the stabilisation of emissions at their 1990 levels by the year 2000, as discussed at the 10th INC meeting in August 1994. A Greenpeace exposé released during the conference revealed that US power generators planned a massive, 20-year construction programme for fossil-fired power plants. This would increase US emissions from electric power generators by 35 percent over the 1990 levels by the year 2014, making the Clinton administration's stabilisation target unrealistic (*ECO*, issue 5, Geneva 1994). The point is that activities in areas where there are strong industrial interests restrict governments' scope for negotiating manoeuvre on the climate change issue.

The European Energy Charter is a further case in point. Here governments produced an agreement that, if implemented, looks certain to counter any measures that the climate negotiation process manages to produce (*ECO*, issue 4 Geneva 1991). The Energy Charter is designed to create Europe-wide opportunities for the expansion of the energy industry[61] and has the strong backing of the energy industries. DGXVII, the energy directorate of the European Commission, which was responsible for the charter, is regarded as 'very responsive to energy industry interests' (Matláry 1991:19). According to environmental NGOs, the European Energy Charter illustrates that 'when push comes to shove, it is the energy lobby inside and outside government which writes the real policies, and not the Foreign or Environment Ministries' (*ECO*, issue 4, Geneva 1991).

This touches on the wider question of interdepartmental power relations and the way in which Trade, Industry and Energy Departments, with which the fossil fuel lobbies enjoy supportive relations,[62] enjoy a privileged position in policy formulation. As Christophe Bourillon of the WCI argued in 1995, 'We have many and wonderful contacts with ministries of industries . . . it is a battle between those ministries at the national level. Our influence probably goes further because the Trade Ministries are more powerful than the Environment Ministries'.[63] The defeat of the carbon tax in Europe (described above) can also be understood in terms of inter-bureaucratic power relations, given that the rejection of the proposal can in part be attributed to the fact that EU Environment Commissioner Carlo Ripa di Meana 'lost his battle with the competition and industry commissioners' (*ENDS Report* 1992d).

These examples draw attention both to the problematic nature of the narrow inter-state conception of the scope of climate politics employed in the regime literature, and to the importance of domestic bureaucratic politics to explanations of the scope and effectiveness of international measures.

---

[61] The charter lays the legal groundwork for East–West cooperation in the business sector on energy issues (Morand-Francis 1995) as part of the EU's energy market liberalisation programme, intended to lower the price of fuel and thereby encourage greater consumption.
[62] Interview with Sale, 18 July 1996.
[63] Interview with Bourillon, 22 June 1995.

5.6   The un-politics of climate change

In many ways, it can be argued that corporate lobbying results not only in an identifiable policy change that bears the hallmark of that lobbying, but also in the absence of policy action on an issue. Policy analysts have referred to this process as non-decision-making (Bachrach and Baratz 1970:7).[64] The term refers to the exclusion of particular issues from the policy agenda, and in this case would apply to policies that governments are unable to consider because of the negative impact they would have on industries represented by the fossil fuel lobbies. Such influence can reveal itself in a number of ways, for example 'when decision-makers ... do not act because they expect opposition from key political actors' (Ham and Hill 1988:65). Crenson (1971) uses the non-decision-making formula to illustrate how air pollution policies in the US were delayed by the reputation of the powerful US steel industry. Crenson observes that, although not politically active, the economic importance of the industry (unsupported by acts of power), was decisive in stalling action. Policymakers' expectation of a negative reaction from key industries was sufficient to delay the formulation of policy on air pollution.

In a slightly different way, the notion of anticipated reaction may said to be at play in contemporary US climate politics. As Hatch (1993:33) notes, 'There would appear to be little inclination to expend precious political capital on global warming by revisiting congressional battles over comprehensive energy legislation, especially in light of the recent brawl in Congress over the energy tax proposal.' The US government knows, in this case based on experience, that the furore and intensity of resistance that would greet any proposal to limit the supply or consumption of energy, make the prospect of legislating in this area undesirable. Moreover Marchetti notes that although industry did not participate in the drafting of the 1994 Italian CO_2 stabilisation plan, 'industry is exerting some influence on the thinking of policy-makers', and 'positional influence will dictate that industry will have the last word on all these actions' (Marchetti 1996:313, 317). Reputation can therefore affect not just the resolution of an issue, but its emergence in the first place.

The power of the lobbies in these situations is perceived *by* governments *of* the corporations. As Bachrach (1967:80) notes, 'once a national issue has come to public attention [companies] frequently exercise their influence whether they want to or not. Their sheer existence owing to their size, power, ubiquity ... leaves them no alternative'.[65] Governments know that the energy industries, upon which they rely, would not tolerate forms of greenhouse abatement action that implied reduced

---

[64] Bachrach and Baratz (1970:7) argue that 'power is also exercised when A devotes his energies to creating or reinforcing societal and political values and institutional practices that limit the scope of the political process to public consideration of only those issues which are comparatively innocuous to A'. Elsewhere they suggest that 'to the extent that A succeeds in doing this, B is prevented from all practical purposes from bringing to the fore any issues that might in their resolution be seriously detrimental to A's set of preferences' (Bachrach and Baratz 1971:379).

[65] Bachrach and Baratz (1970) refer to this as the rule of anticipated reaction.

consumption or production of energy. Hatch (1993:29) argues that the US government is 'well aware of the powerful constituencies affected by the large distribution of income that would result from ... interventions in the economy that would restructure energy production and consumption patterns'. This encourages decision-makers to factor in to their evaluation of appropriate policy, the anticipated response and perceived preferences of important constituencies, conferring a great deal of tacit influence and latent power upon corporate actors. These, then, are examples of the non-decision-making scenarios that attend global warming, areas of what Crenson (1971:184) calls 'politically enforced neglect'.

Moreover the absence of a lobbying organisation for a certain sector can be evidence of the extent of its representation within government, and the lack of countervailing challenges to its position of strength. An extreme example of this is the case of the OPEC states, where there is no need for oil lobbies to press their interests on governments whose stability rests on pursuing the same interest. In Bourillon's words, 'OPEC states don't need any lobbying from business'.[66] A similar situation could be said to prevail in Japan, where groups such as the Electrical Manufacturing Federation and the Automobile Industry Association go largely unchallenged in the absence of significant environmentalist activity (Wilson 1990). Where there are no counterchallenges or interest groups attempting to push policy in a different direction, political mobilisation becomes less imperative.

These examples seem to suggest that visible participation in the policy process and manifestations of conflict between interests offer an incomplete guide to political influence, and that areas of political neglect may also be indicative of the presence of influence.[67] Power does not have to be exercised to be present, and whether or not power is exercised in a conscious manner is irrelevant to the question here, of how actors influence policy outcomes. The most important argument in relation to this chapter, is that the input of the fossil fuel lobbies into the policy debate on global warming is privileged. It is not that industries do not need to lobby to help secure their interests, for they will of course take every opportunity to make their voice heard in seeking to steer the course of policy. In any case, their expertise is regarded as being instrumental to policy formulation. The issue is the way in which governments more readily internalise, and hence adapt to, their concerns than is the case for other groups, because the goals of government and the fossil fuel industries coincide in important ways with respect to climate change. Almost by definition, those states whose responsibility it is to activate a response to global warming, are those states in which the economy and the energy resource base are closely related;[68] those states, therefore, in which political lobbies organised around these resources

---

[66] Interview with Bourillon, 22 June 1995.

[67] These ideas have much in common with a Foucauldian analysis of power (Foucault 1980), where the very invisibility of power is a testament to its success, and with Susan Strange's use of the term structural power to denote the power to control the context within which others make their decisions (Strange 1988). See Chapter 2 for more on this.

[68] Because these are the states with the highest output of energy-based emissions.

are most prominent and influential, and most in line with the government position on the issue.

Building on a similar argument, a case can be made that the national interest is conceived in broad economic terms, so that defence of the interests of key industries is regarded as synonymous with protection of the state's interests (Gill and Law 1988; Jessop 1990).[69] This is perhaps especially the case with energy policy, because it is closely linked to economic growth and therefore touches on a core state interest (Newell and Paterson 1998).[70] The lobbies are therefore able to present their interests as those of capital-in-general. According to World Coal Institute lobbyist Christophe Bourillon, 'What governments want is to stay in power, and to stay in power they have to please voters, and to please voters there have to be jobs and growth, and in that sense the way governments go is the way we go because development is the goal'.[71] It is also useful to note, as Paterson (1995b:4) does, that 'The state is extensively involved in the energy business in most capitalist countries' through ownership, funding the development and protection of markets. Such involvement can be argued to strengthen the mutual dependency between the state and the energy industries.

An argument about the privileged position of the lobbies in the policy process helps to make sense of the empirical story above. Whereas direct lobbying can visibly alter the course of initiatives or frustrate their implementation, indirect and tacit forms of power reveal themselves in the suppression of consideration of policies that are damaging to perceived interests, and restriction of the debate on appropriate responses to the threat of climate change to gesture politics and the sorts of tokenistic ventures that have come to feature highly in government programmes post-Rio. In both instances, the fossil fuel lobbies are able to shape, to varying degrees, the contours of international cooperation on global warming.

5.7 Conclusion

The sets of relations described above are of course subject to change. An account such as this can only ever provide a snapshot of the networks of influence, of which the fossil fuel lobbies are a part. Changes within the lobbies were apparent at COP1, where hardline US coal and oil lobbyists found some of their traditional allies, concerned about losing the goodwill they had developed with the states earlier in UNCED process, choosing to distance themselves from the more confrontational

---

[69] The way in which the British government defended its opposition to the proposed EU carbon tax – by utilising arguments provided by industry in defence of the fossil fuel sector (*ENDS Report* 1992c; Rowlands 1995:142–3) – suggests another example of the way in which industry's assumptions about the threat to competitiveness posed by the tax have been adopted by politicians.

[70] Dahl (1994) argues that governments' sensitivity to energy questions is also borne out by the fact that the Maastricht Treaty on European Union permits majority voting on environmental questions but not on energy policy, which has to be made on a unanimous basis.

[71] Interview with Bourillon, 22 June 1995.

elements of the lobby. Even the oil lobby, with obvious interests to defend in the debate, was divided between those with interests in gas[72] and those firmly allied to the stance of their counterparts in the coal lobby against further international action (Grubb 1995b). Since that time, BP and Shell have both parted company with the GCC (but not the American Petroleum Institute) as a result both of increased investment in renewables (BP 1998) and photovoltaics, and of the perceived counterproductive effects of GCC's aggressive lobbying stance, which, whilst suited to the Washington policy style, was making little headway at the UN, and was out of step with the attitude of European governments with which Shell and BP have to cooperate. Michael Brand, Shell's senior environmental policy advisor, was said to have been appalled at the 'out and out Congressional lobbying set up by the GCC' (Boyle 1998). Hence while GCC led opposition to US ratification of the Kyoto agreement, Shell declared its support for the Kyoto process in European fora. In an even more surprising turn of events, Shell joined the Wind Energy Association in order to promote the 5–10 percent market in wind power it hopes to capture by 2010.

This is also indicative of a general difference in outlook between US and European industries: the latter are more resigned to the inevitability of climate action and are keen to forge a role for themselves in playing a constructive part in the policy debate, while the former, because of the state of play in the climate debate in the US, have been able to centre on whether or not climate change is a problem at all. In Europe the debate has moved on to which response is more appropriate. No major European industry federation is formally opposed to the Kyoto Protocol. Some European oil companies, such as OMV, have even gone as far as to support the EU proposal for a 15 percent reduction in CO_2 emissions (Greenpeace International 1998:52).

Changes in the relationship between the lobbies and different government administrations can also be expected to change the landscape of climate politics, as the current dominance of the US Congress by Republicans, anxious to turn back the tide of environmental legislation, bears out. Kinrade also argues that 'influence on climate politics has swung very much in favour of the resource industries in Australia in the last couple of years,[73] a shift that has been consolidated under the Liberal party administration of John Howard. This was in evidence at COP2, where the Australian delegation's position was firmly against further action and increasingly close to the obstructionist ground traditionally occupied by the OPEC states.

The picture that emerges from the above discussion, however, is that proposals for emission reductions in the future will be resisted by powerful lobbies with a

---

[72] Some former members of the GCC, particularly gas companies such as the American Gas Association, have become members of the Business Council for a Sustainable Energy Future (Pearce 1995d). Insurance companies have also told the ICC to drop its anti-abatement line or lose their support (Brown 1996). America's biggest gas company, Enron, and Daimler Benz also produced briefings entitled 'Yes to Kyoto' in the run-up to COP3 (Leggett 1997:5).

[73] Questionnaire from Kinrade (1996).

substantial capacity to influence state decision-makers. This suggests the need to temper expectations about likely solutions to the issue of climate change, by reference to the constraints that inhibit states' ability to select policy strategies. This analysis would seem to suggest that any assessment of what it is realistic to expect from near-term climate negotiations, needs to incorporate an assessment of the interests aligned against further action. As Hatch (1993:7) puts it, 'given the broad sets of interests activated by global warming concerns and the ready access those interests have to decision-making bodies . . . the ability of government to obligate itself internationally to measures requiring positive actions domestically is severely circumscribed'. The fact that the negotiations witnessed governments anxiously protecting the very industries that contribute on a large scale to the greenhouse effect, brings into focus the problem of expecting states to regulate sectors of industry with which they share key interests. This is particularly so, given that those countries from which action is most required are also states where energy use is central to the economy, and therefore in which the most vocal lobbies often reside.

Hence whilst it is difficult to generalise, given that (as the above examples illustrate) there are circumstances in which the power of the lobbies is reduced and particular government set-ups inhibit interest group influence, it does seem that an account of global climate politics that overlooks the importance of the fossil fuel lobbies would be lacking in many respects.[74] At minimum, the paralysis that characterises the efforts of the international community to confront climate change cannot be understood in isolation from the power and influence of the fossil fuel lobbies. Strictly inter-state accounts of international cooperation miss many of the important global interactions described above, where groups pressure governments domestically to adopt policies that are conducive to their interests internationally, and in turn governments are constrained in international negotiations by the coalitions of interests upon which they rely for support domestically. Hence it is not just a question of states' ability to cooperate internationally being curtailed at the national level. The above examples also demonstrate how the ability of states to act in the domestic sphere is restricted by the manoeuvrings of global lobbies.

It has not been suggested that corporations are more powerful than states in the international politics of global warming (as if they could in any case be meaningfully compared). States remain the key actors in many senses. States act as gatekeepers of demands, respecting some and neglecting others. What is notable is that the way in which governments filter demands is favourable to particular actors and agendas.

---

[74] For a comparative account of different corporate–state relationships and the way in which this affects environmental policy see Aguilar (1993). See Wilson (1990) for a comparative history of relations between businesses and the state in different countries.

6

Climate for change: environmental NGOs

6.1 Introduction

This chapter looks at the political influence of environmental pressure groups on the way in which the issue of climate change has been addressed at the international level. The focus here is essentially advocacy organisations, and not primarily research or think-tank oriented NGOs,[1] despite a degree of cross-over between the two.[2] The scope of analysis is further narrowed by a focus on groups that have a key interest in global warming and are active at the international level on the issue.[3]

Despite an emerging literature on global environmental politics that acknowledges, in passing, the increasing part played by environmental non-governmental actors in the resolution of global environmental problems (more often than not through brief reference to the astronomical rise in their number and size in recent years), relatively little has been written on the actual impact of the groups upon global policy, besides vague and unexplored references to their centrality in the development of international environmental policy.

The literature on environmental NGOs can be divided into three broad categories: (1) Potter (1996a, 1996b) and Willetts (1982, 1996c) examine the institutional roles performed by NGOs in a case study format; (2) Chatterjee and Finger (1994) and Finger (1993) provide a critical perspective on the role of NGOs at the global level, centring on the issue of co-option by global 'power brokers' and (3) Wapner (1996) and Princen and Finger (1994) look at the ways in which NGOs transform the *nature* of world politics. The relative absence of analyses of the impact of environmental NGOs upon outcomes at the international level, extends to global climate change, where the *influence* of environmental pressure groups has been neglected (with the exception of Arts 1998, whose work is discussed later in the chapter) and their *role* in more general terms is only sparsely covered (Dubash and Oppenheimer 1992; Rahman and Roncerel 1994). Hence this chapter seeks to explore the extent to which the political influence of environmental NGOs helps to account for global climate politics.

---

[1] For different categorisations of environmental NGOs see Caldwell (1990), Elliott (1992) and Willetts (1993).

[2] Some of the groups looked at here, for example, are affiliated to think-tanks. The World Resources Institute, for example, has representatives from WWF (World Wide Fund for Nature), NRDC (Natural Resources Defence Council) and EDF (Environmental Defence Fund) on its Policy Panel on Responses to the Greenhouse Effect and Global Change (WRI 1991).

[3] Conca (1995:444) estimates that 'only a relative handful of environmental NGOs are truly transnational'.

Part of the problem in attempting such an assessment is that drawing on accounts provided either by environmental pressure groups themselves, or the government officials they seek to influence, provides an incomplete and potentially distorted representation of the nature and extent of their influence. This is so because pressure groups may be inclined to exaggerate the degree of influence they have been able to exert, and governments are as likely to claim NGO successes for themselves, or to recast events in a light that reflects well on their role (Elliott 1992).

This is complicated by Barratt's (1996:9) observation in relation to the political influence of environmental groups (but applicable more widely), that 'Nobody can know for sure that their advice was the turning point over some issue, so the impression of having an influence is never confirmed.' Classic definitions of influence employed in the literature on environmental NGOs, where A intentionally transmits information to target organisation B that alters B's policies (Potter 1996a, 1996b), are limiting, for as Humphreys (1996:102) notes, 'Due to the frequently informal and invisible nature of the interactions between NGO representatives and government delegates, assessing the impact of NGO activity is not an easy task.' Here Potter (1996b:49) concedes that researching NGO influence requires 'making informal judgements based on incomplete evidence'. The analysis of influence that follows, combines an agent-centred account of the operation of the groups in direct lobbying terms, with arguments about the bargaining assets of actors, implying a more structural perspective. The analysis, therefore, goes beyond the observable interactions between governments and environmental pressure groups, and explores less observable influences on global climate politics through ideas about anticipated reactions and tacit forms of influence.

This approach complements the work of Arts (1998), which focuses strictly on the observable and intentional influence of environmental NGOs upon various issues in the international climate negotiations, traced through NGOs' own perceptions of their influence (what Arts refers to as ego-perception), government officials' perceptions of the impact of environmental groups (termed alter-perception) and attempts at tracking events through causal analysis. The limits of this more positivist analysis are acknowledged by Arts, who argues that a fuller account of influence must seek to explain lack of influence, as well as develop counterfactual analysis and look at the connections NGOs are able to forge between national and international politics (ibid.:321).

6.2   Global warming and environmental pressure groups

The issue of global climate change presents environmental NGOs with a particularly difficult set of obstacles and challenges. Besides the very many scientific uncertainties associated with global warming, which make it harder to develop a consensus on appropriate policy action (Skolnikoff 1990), there are also the economic costs that are perceived to be involved, and the scale of economic restructuring (in terms of energy production and consumption) that may be implied by efforts to reduce greenhouse gas emissions. The interests that will be affected by such change are amongst the most powerful in the global economy (see Chapter 5). In this respect, Conca (1995:454) argues that environmental NGOs' 'resources and access to political power pale in comparison to the forces driving environmental destruction'. It comes as no surprise, then, that environmental NGOs consider that the strength of the interests working to roll back the gains they are campaigning for inhibits their ability to advance their agenda on global warming.[4] There is also a long time lag between the point at which action is taken to reduce the onset of global warming and the observable effects of such action. This disinclines governments to take potentially very costly action, the benefits of which will be not be seen for many decades and for which they will not receive credit.

The relative absence of convenient 'techno-fixes' with which to combat the problem, closes further potential channels of influence in terms of group advocacy of simple solutions to the problem.[5] It is easy to contrast this scenario with the case of ozone depletion, where consumer boycotts have been an effective in hastening the elimination of CFC (chlorofluorocarbon) production (Bramble and Porter 1992; Litfin 1994; Rowlands 1995) and calls for the abolition of ozone-damaging chemicals were realistic. In relation to ozone, the issue was replacement and substitution. With regard to climate change, the issue is arguably 'dissipating' business and not 'different' business (Rowlands 1995:137). Hence, besides trying to 'highlight the commercial opportunities for innovative technologies that will emerge within the framework of an international convention' (Dubash and Oppenheimer 1992:276), environmentalists have to lobby for tough policy choices that will affect status quo stakeholders.

Making connections between distant global processes and the effect of everyday actions, which is problematic for any group campaigning on global environmental issues, is particularly difficult in the case of global climate change, where the effects are uncertain and the sources of the problem so disparate. Exploiting the mass media in making these connections has also been regarded as difficult by many environmental NGOs (see Chapter 4). NGOs believe that a great deal of their influence depends on positive media coverage, which they feel has not been forthcoming on climate change.[6] This may relate to the fact that, unlike many other international environmental issues, global warming is less easily presented in terms of convenient and strategically useful oppositions of 'right' and 'wrong', in a way that debates on whaling or acid rain can be. With these latter issues, victims and perpetrators are more readily identified than is the case with global warming, where responsibility for causing the problem is more clouded and diffuse. The fact that those states which contribute most to climate change are not the states that are likely to be most

---

[4] Questionnaires from Weir (1996), Spencer (1996), Leggett (1996), Kinrade (1996).

[5] This is not to say that groups have not attempted to emphasise the use of climate-benign and energy-efficient technologies. Greenpeace, for example, has exposed those retail stores that have not accepted its climate-friendly 'Greenfreeze' refrigeration system (Greenpeace Business 1994b).

[6] Questionnaires from Weir (1996) and Kinrade (1996).

affected by its impacts, makes it harder to convince governments to take responsibility for their actions as few immediate gains will accrue to them directly.

Moreover the very different agendas that pressure groups from North and South
bring to the issue of climate change, complicates the formation of North–South
NGO alliances that might improve the prospect of international action. Related to
this is the fact that thorny issues of equity, consumption and rights to development
feature highly in the global warming debate, making it harder for environmental
pressure groups to bring about change when hugely conflictual and divergent state
agendas are at play.

These factors help to explain why environmental NGOs believe that most governments, particularly the largest emitters of greenhouse gases, have shown a lack
of political will to act on the issue of climate change. As Weir notes, 'very few governments are prepared to take any but small incremental steps towards solving the problem'.[7] It needs to be acknowledged at the outset, therefore, that global warming
brings its own set of unique and particularly perplexing challenges to environmental
NGOs seeking to encourage international action. The question of how far the problem structure of global warming inhibits significant NGO influence will be
returned to at the end of the chapter.

6.3   Scope of the analysis

The Climate Action Network (CAN) enjoys the membership of most international
NGOs that are active on climate change, and will therefore be the main case study
in this chapter.[8] CAN was created in 1989, prior to the Second World Climate
Conference (SWCC) by sixty-three NGOs from twenty-two countries under the
initial guidance of groups such as Environmental Defense Fund (EDF) and
Greenpeace International.[9]

The major focus of CAN's campaigning is the contribution that industrialised
countries make to the problem of global warming. This has taken the form of directly
applying pressure on Northern states to reduce their emissions of greenhouse gases.
CAN has attempted to provide a coherent NGO voice from the earliest stages
of discussion on the climate issue. It is regarded by its members as performing
an indispensable role in the coordination of strategy and campaigning activity, by

---

[7]   Questionnaire from Weir (1996).

[8]   For a clear articulation of the policy position of CAN on the climate issue see the statement by environmental
NGOs present at INC1, (Intergovernmental Negotiating Committee for a Framework Convention on
Climate Change), reproduced in *ECO*, issue 1, Chantilly 1991. CAN's initial goal was to press for a reduction
of at least 20 percent in the 1988 level of CO_2 emissions by the year 2005 (Porter and Brown 1991), based on
the recommendation of the Toronto Conference on the Changing Atmosphere. For the evolution of NGO
positions during the course of the negotiations see Rahman and Roncerel (1994).

[9]   Originally the network only included NGOs from the industrialised world, principally Western Europe,
North America and Australia, but it has now taken on members from Asia, Africa, Latin America and Eastern
Europe and links more than 100 groups around the world (Rucht 1993).

orchestrating common positions among NGOs and keeping them informed of the latest developments in climate policy debates.[10]

6.4   Structural factors/bargaining assets

This section briefly discusses a number of positional factors that affect the ability of environmental groups to define the international agenda on global warming. The assets that groups may be said to possess and the restraints on their ability to exercise influence are discussed here at a high level of generality. Specific contexts are explored in more detail in the subsequent sections.

One factor that militates against environmental groups is the ineffectiveness of the institutions set up to coordinate international environmental action, such as the United Nations Environment Programme (UNEP), which is both under-resourced and relatively powerless in the overall structure of the United Nations machinery (Conca 1993, 1995; McCormick 1993). Given also that the United Nations Conference on Environment and Development (UNCED) process encouraged groups to formulate common positions and suppress differences of opinion (Chatterjee and Finger 1994), the different perspectives that the diverse groups campaigning on the climate issue bring to bear at international meetings do not necessarily filter through to policy-makers.

Despite the fact that there has been close coordination of the efforts of all the major groups campaigning on the issue of climate change,[11] accommodating the perspectives of so many groups and coordinating effective campaigns amid this diversity is ridden with problems that impinge on the overall influence of a coalition. One way in which CAN has coped with this has been to create regional climate networks.[12] This is particularly important for Weir, who notes that 'global warming is too large an issue for campaigning to be effective unless there is simultaneous pressure at all levels'.[13] Regional and international coalitions have been able to strengthen the position of NGOs campaigning primarily at the national level (Hawkins 1993; Porter and Brown 1991).[14] The breadth of concerns captured by such a coalition can also be regarded as a positive asset. CAN contains a wealth of diverse expertise: World Wide Fund for Nature's (WWF) climate change campaign has focused on the negative effects on biodiversity (Markham 1994; WWF 1996); Greenpeace has sought to emphasise the human impacts of climate change (1994a); and the EDF's global

---

[10] Questionnaire from Weir (1996).

[11] Questionnaires from Sharma (1996), Leggett (1996), Weir (1996), Kinrade (1996), Spencer (1996), Stanford (1996), Huq (1996).

[12] For example CANLA (Latin America) CNA (Africa) CANSA (South Asia) CANSEA (South-East Asia) and CNE (Climate Network Europe).

[13] Questionnaire from Weir (1996).

[14] For an overview of the needs, forms and problems of cross-national cooperation among environmental groups see Rucht (1993).

atmospheric programme boasts the largest assemblage of scientists, economists and lawyers of any national NGO working on climate change.[15]

Unlike its industry counterparts such as the World Coal Institute and the Global Climate Coalition, the CAN network does not have a permanent institutional base. The CAN coordinator is an appointed member of one of the member groups with the resources and back-up to perform the function. Moreover there are very few groups campaigning exclusively on climate change, and those that do exist are poorly resourced (for example the Global Commons Institute). CAN enjoys a collective global membership of twenty million people (*ECO*, issue 6, August 1994). This forms the basis of CAN's claim to speak for a constituency beyond its own organisation alone, to represent the public interest to a greater degree than other non-state actors.

In the sections below, the political impact of environmental NGOs is considered at different stages of the policy process, from agenda-setting through to implementation.

6.5 Agenda-setting

Rahman and Roncerel (1994) argue that environmental NGOs enjoy a great deal of influence in national policy arenas, where policy responses develop first. Their argument is reinforced by a survey carried out by the International Institute for Applied Systems Analysis, which discovered that most NGOs consider personal contact with politicians early in the policy process, to be the most effective channel of influence (ibid.).[16] The lack of efficiency and effectiveness of international environmental institutions (see above) also means that groups show a preference for national-level lobbying, even after the issue has begun to be dealt with in international fora.

It is at this stage of the policy process – when the problem is defined, expertise sought and the need for international action discussed – that policy positions are developed. McCormick (1993) shows how governments often react to catalysts rather than adopt proactive positions, creating opportunities to define issues, shape responses and force the agenda. In order to encourage governments to act on an issue imbued with complexity and affecting a range of powerful interests, the generation of a public consensus in favour of action (or at least the perception of a public consensus) can be instrumental in helping to set political agendas (Benedick 1991a). Environmental pressure groups can play a pivotal role in the activation and vocalisation of this popular concern. One of the greatest impediments to the resolution of a complex political problem such as global warming is social inertia (Dubash and

---

[15] The Natural Resources Defence Council has adopted a similar approach in its campaigning on climate change, employing scientists and lawyers to lobby for changes in US policies and to bring law suits. See Lashof (1993).

[16] Boyer's (1997) work on NGOs and Swiss climate policy shows that the influence of NGOs such as Greenpeace and WWF-Switzerland is principally exercised through national level structures and processes, despite participating in international negotiating teams.

Oppenheimer 1992:278). The generation of focused demands, on the other hand, can smooth the passage of an issue from the political margins to political centre-stage.

Groups' ability to set the agenda also depends on their ability to raise an issue circulating within the scientific community to the status of a high profile political issue. NGOs are able to draw attention to reports and studies that have been 'conveniently neglected by governments' (Stairs and Taylor 1992:112).[17] They are able to politicise an issue that was not previously considered political in an overt sense, and in so doing they 'play a critical role in minimising the time from recognising the problem to setting the agenda' (Breyman 1993:138). Environmental NGOs are therefore able to 'create' issues and to push them onto institutional agendas.[18] Lashof argues more specifically that environmental NGOs are most effective through their ability to 'translate knowledge of what is going on to public outrage, that results in publicity at key times when decisions are being made'.[19]

In this sense NGOs can call to bear 'considerable if diffuse public support' (Hawkins 1993:222) for goals such as the negotiation of a climate convention. As a UK government representative at the climate negotiations, Philip Dykins, argues, 'You can't get politicians to take tough decisions without popular backing.'[20] Weir supports this argument by noting that contact with diplomats at international meetings, lobbying at the national level and cooperation with corporate actors are 'only effective when backed by public pressure'.[21] The ability of environmental NGOs to appeal to a broad constituency of support for their goals, in many ways appears to preclude their use of other channels of influence.[22]

Particularly during periods of rapid growth in group membership, pressure groups are able to emphasise the extent to which there is public concern about a particular issue and an institutional response is expected. Sebenius (1991:122), for example, attributes 'The significant number of industrial countries that unilaterally, or in small groups, had committed [themselves] by late 1990 to greenhouse gas stabilisation or reduction targets' to the 'high level of public concern about the greenhouse issue.' This function can also take the form of supporting moves within government to adopt a more forthright line. Lashof argues that there are often groups within government pushing for action, 'but their ability to move things forward depends on whether or not there is a constituency for doing that, and so the

---

[17] Interview with Silberschmidt, 18 July 1996, Geneva.

[18] This has resonance with the social movements literature, which finds that 'social movements bring up new issues, politicise them within civil society and prepare the grounds for the political system to integrate them' (Finger 1994:32). Litfin (1993:95) notes, moreover, that 'social movements have instigated virtually all existing international environmental agreements'.

[19] Interview with Lashof, 18 July 1996, Geneva.

[20] Interview with Dykins, 25 February 1994, London.

[21] Questionnaire from Weir (1996).

[22] It is clear that many groups regard demonstrating public concern as their most important function (telephone interview with Robertson, 28 April 1994) and generating public pressure as their most effective channel of influence. (Questionnaires from Sharma (1996), Stanford (1996), Weir (1996).

role of NGOs is doing things which the administration simply cannot'.[23] Creating a substantial constituency of popular concern requires getting publicity for an issue, mobilising other NGOs and lobbying legislatures. This creates a supportive environment for governments to take a position that may otherwise be considered too risky.[24]

Following on from this, NGOs can also create a sense of public expectation about the sorts of policy response that are deemed desirable. Pressure on the US to be party to the Framework Convention from 1991 onwards can be seen in this light, as the expectation of positive participation by the US in the global climate negotiations was widespread (Brenton 1994).[25] Flagging political will in 1991 on the part of the US administration was the target of a fierce NGO and media backlash, which was successful in bringing about a renewed government commitment to adopt a more proactive line. Lashof argues, 'in the run up to Rio, the pressure from around the country, but led by some of the national groups such as NRDC and EDF, certainly made a big difference in convincing President Bush that he had to go to Rio and he had to have a legally binding treaty'.[26] More recently, CAN has also been working to shame Australia and New Zealand for their failure to support the negotiating position of their Pacific neighbours in the climate negotiations (Arts and Rüdig 1995). Mintzer and Leonard (1994:36) note in this respect that NGOs can play a critical role in 'focusing public attention on outstanding issues or outrageous national positions at critical moments in the process'.[27]

By nurturing frames of interpretation supportive of their position, NGOs can have some influence on the terms of debate. When an issue explodes onto the political scene, as the issue of global warming did so suddenly in 1988 amid intense public concern about dramatic changes in the weather (Ungar 1992), governments are expected to react with a series of policy proposals. In a reactive and turbulent atmosphere such as this, NGOs are able to set the pace of political change and assert their preferred interpretations of what the issue means and what degree of action may be appropriate, whilst governments are deciphering their own preferences and interests in the debate. Princen et al. (1995:52) describe this as 'setting the conditions under which states will act or, maybe more precisely, react'. NGOs can serve as 'scientific advisers or information gatherers at the point at which risks are being defined or problems being diagnosed' (Susskind 1994:49), and thus influence how interests come to be understood in relation to the issue. When governments lack the required interpretive packages to respond to a new problem, environmental NGOs can perform the sort of agenda-setting function that Haas (1990a) attributes to knowledge-based epistemic communities. They do this by infusing a new issue

[23] Interview with Lashof, 18 July 1996, Geneva.
[24] Ibid.
[25] Interview with Lashof, 18 July 1996, Geneva.
[26] Ibid.
[27] Interview with Sieghart, 17 July 1996, Geneva.

with meaning, scale and urgency in a way that resonates with existing popular concern.

In this context, NGOs riding a wave of popular concern are often able to extract promises of action, however tokenistic or incremental in nature. This can be seen in the US, for example, where the government's recognition of public alarm about the issue of climate change, together with popular concern about the lack of political action, encouraged the announcement of a $1 billion fund for climate research, a plan to plant a billion trees a year and the government's intention to host a White House Conference on climate change in the spring of 1990 (Brenton 1994).

Further evidence of agenda-setting by environmental groups was the organisation by the Stockholm Environment Institute of the Villach and Bellagio workshops in 1985, which set the pace of the climate change debate at the time, prompting governments to develop their own institutional response in order to ensure that they would have control over the future development of the issue (as opposed to being forced to respond to NGO initiatives). One such institutional response was the establishment of the Intergovernmental Panel on Climate Change (IPCC) to provide a scientific basis for the political negotiations on climate. By helping to nurture an embryonic consensus among scientists that action of a precautionary nature was desirable, NGOs were able to force states to respond to their initiatives and quicken the pace of political action on the issue. As William Nitze, former deputy assistant secretary for the environment in the US, acknowledged, 'the two workshops ... indeed played a significant catalytic role in establishing the IPCC ... governments could no longer permit ... NGOs to drive the agenda on the emerging climate issue' (quoted in Dubash and Oppenheimer 1992:265).

In some cases, however, NGOs merely react to government initiatives on climate change. This was the case in Germany for example, where NGOs were initially reluctant to take up the global warming issue for fear of playing into the hands of the nuclear industry, whose agenda they strongly opposed.[28] Similarly, the governments of many Southern countries claim to have sought the involvement of NGOs in climate policy development, but found them to be unresponsive.[29]

One key determinant of the degree of influence that an NGO is able to exert over policy is the closeness of its relationship to principal decision-making bodies and actors. Some groups have been centrally involved in the evolution of climate policy at the national level. EDF helped to draft the US government's Carbon Dioxide Offsets Policy Efficiency Act of 1991. Many NGOs have a high degree of access in the form of direct consultative meetings with governments and regular meetings with senior civil servants.[30] CAN-UK members, for example, enjoyed frequent meetings with the UK negotiating team to the Intergovernmental Negotiating Committee (INC) and with the Secretary of State for the Environment, before major

---

[28] For more on German environmental NGOs and the issue of global warming see Cavender and Jaeger (1993) and Hatch (1995).

[29] Interviews in Geneva with Ratnasiri on 16 July 1996 and Rahman on 19 July 1996.

[30] Questionnaires from Spencer (1996), Stanford (1996), Kinrade (1996).

negotiating sessions at the international level, including INC meetings and meetings of the EU Council of Ministers.[31] NGOs, particularly those with a more strictly research orientation, are requested to contribute to and comment on the preparation of government positions. In the US, World Resources Institute (WRI), NRDC, EDF, the Union of Concerned Scientists, the Woods Hole Research Center and the Audubon Society all worked closely with US policy-makers and UN agencies in formulating policy options on climate change.[32] In the South, the Tata Energy Research Institute (TERI)[33] and the Centre for Science and the Environment (CSE) developed policy options for India. The CSE and TERI are said to have exerted a 'crucial influence' on the initial policy-making of the Indian government in 1990–2. 'Typically the two organisations would analyse the latest aspects of the international negotiations and their importance in a briefing to the MoEF (Indian Ministry of Environment and Forests). The ministry simply came to use TERI and CSE as their expert information units and crucial discussion partners in the shaping of a position on climate change' (Jakobsen 1997:8). Ravi Sharma, associate director of the Centre for Science and Environment (CSE) in India, argues that through direct consultative meetings with government they have obtained a high degree of access.[34]

The Bangladesh Centre for Advanced Studies (BCAS) has played a key role in developing their governments' climate policy.[35] Huq describes BCAS as having a high degree of access to government through direct consultative meetings and participation in policy committees, and the overall influence of BCAS in this respect as 'great'.[36] According to Rahman and Roncerel (1994) and Reazzudin, deputy director of the Environment Department in Bangladesh, work by BCAS on sea-level rise helped to convince the Bangladeshi government of the threat that climate change poses to coastal populations.[37] BCAS is represented on the inter-ministerial committee that formulates policy on climate change. Reazzudin says of the relationship between BCAS and the Bangladeshi government, 'we are partners'.[38] In Africa, Environment and Development Third World (ENDA) played a 'pioneering role' in developing emissions inventories (Rahman and Roncerel 1994). Given the gaps in LDC governments' expertise on the issue of climate change and their consequent need to rely on the expertise of NGOs, Rahman argues that 'we [the NGOs] are the government in 90 percent of cases'.[39]

These examples reinforce Banuri's (1993) point that NGOs' ability to influence their governments depends on the governments' access to other sources of research

---

[31] Questionnaire from Weir (1996).
[32] For more on these groups, loosely defined as part of the 'Big 10' US groups, see Mitchell (1991).
[33] See Achanta (1993) and Pachauri and Bhanari (undated).
[34] Questionnaire from Sharma (1996).
[35] Interview with Rahman, 19 July 1996, Geneva.
[36] Questionnaire from Huq (1996).
[37] Interview with Reazuddin, 17 July 1996, Geneva.
[38] Ibid.
[39] Interview with Rahman, 19 July 1996, Geneva.

and information, and, in broader terms, whether NGOs have assets that governments think they can make use of. Hence, by way of counterbalance to the notion that NGOs can set frames of reference, project interpretations and trade their expertise for access to governments, it should be acknowledged that the nature of the relationship between pressure groups and governments and the scope for NGO input is highly dependent upon the willingness, or otherwise, of governments to open up the policy development process to their participation. However, where NGO research is used by governments, NGO frames of reference are imported into government decision-making, thereby extending the influence of these groups.

Some governments have been more positive than others in encouraging NGO participation in policy formulation. Bill Hare, climate campaigner at Greenpeace International and formerly of the Australian Conservation Foundation, praised the 'unusually comprehensive' approach adopted by the Australian government, which allowed NGO representatives open access to the Greenhouse Coordinating Group during the development of its Ecologically Sustainable Development Strategy, in which action on climate featured highly (*ECO*, issue 6, Geneva 1991; Russell 1994).[40]

Participation in policy deliberations does not mean that groups' concerns are necessarily acted upon, however. The US provides a useful example. Despite having one of the more open greenhouse policy development processes, in which environmental pressure groups have participated extensively, US policy towards global warming remains one of the most backward of all the OECD countries. The influence of the fossil fuel lobby groups (see Chapter 5) and the presence of White House Chief of Staff John Sununu, are among the factors that explain why the US government has been able to stall on more stringent action to combat climate change, and to smother calls for further action by environmental pressure groups. The hostility of the White House Chief of Staff to environmentalists was made clear by the way in which he dismissed their concern about global warming as part of a hidden agenda to prevent growth in the US economy. He is quoted as saying, 'Some people are less concerned about climate change than establishing an anti-growth policy' (*ECO*, issue 4, Nairobi 1991).[41]

Other states have actively tried to reduce, if not marginalise altogether, the degree of NGO input into policy debates. Egypt, for instance, suggested the elimination of any reference in the convention text to cooperation with NGOs in policy debates (*ECO*, issue 6, New York 1992), while Saudi Arabia, Argentina and Mexico expressed concern about French proposals to increase the degree of consideration given to NGO information by the Conference of the Parties (*ECO*, issue 3, Geneva 1991). The very different attitudes of states towards NGO participation undoubtedly condition the ability of NGOs to influence their own government's position at

[40] Questionnaire from Kinrade (1996).
[41] McCormick (1991:46) shows how, despite regular contacts with ministers and government departments, environmental groups in the UK have 'not been able to translate numerical power into appreciable and consistent direct political influence of the kind enjoyed by economic interests.'

the national (as well as the international) level. The degree to which there are democratic channels within governments for NGO pressure, is a further important factor. Singh notes, for example, that the greatest obstacle to CANSEA influencing the position of the Indonesian government is the 'limited democratic space in which we can work'.[42]

The relationship between groups and the administration they are seeking to influence is also, of course, in a continual state of flux, with different governments being more or less sensitive to the demands made of them by environmentalists. While the Bush and Reagan governments in the US were hostile to action on climate change, close ties between government departments and US environmental groups in the 1970s under the Carter administration allowed for the funding of pioneering climate change research (Bramble and Porter 1992:321; Mitchell 1984).[43] The Clinton administration is also regarded as being more sympathetic to the views of environmentalists on the issue.[44] The presence of political parties such as Die Grünen in Germany (in government in this case) can also allow for more cooperative relations to develop between the government and NGOs, where parties are more receptive towards NGOs' concerns. As Dubash and Oppenheimer (1992:271–2) note, 'the current strong position of the German government (relative to the US for instance) on ... global warming bear witness to ... the success of the electorally partisan approach in the European context'. Hence whilst North American NGOs may have been successful in influencing particular policies on an issue-by-issue basis, they lack the political base that is afforded to some European environmental campaigners.[45]

NGO influence is contingent upon groups' relationships not only with particular governments or parties, but also with particular government departments, and how influential those departments are in the overall policy process.[46] Peter Kinrade of the Australian Conservation Foundation argues that 'some ministries (e.g. environment) are very accessible. Others in key programme areas (e.g. energy and transport) less so.'[47] Japanese climate campaigner Masatabe Uezono argues in a similar vein, that 'MITI [Ministry of International Trade and Industry] and industry groups are very close and that is the biggest obstacle for groups like CASA' (Citizen's Alliance for Saving the Atmosphere and Earth).[48] Hence, whilst NGOs

---

[42] Interview with Singh, 17 July 1996, Geneva. See Potter (1996b) for more on the way in which democratic structures can affect NGOs' influence.

[43] See also McCormick (1991) on the importance of different governments in terms of their openness to dialogue with environmental pressure groups in the context of the UK.

[44] Interview with Reinstein, 18 July 1996, Geneva.

[45] A case can be made, however, that the existence of a Green party is not always a strengthening factor. Chatterjee and Finger (1994:67) note that 'the success of Green parties has substantially weakened most other environmental agents in Western Europe'. See also Hatch (1995).

[46] Interviews with Robertson on 28 April 1994 (telephone), Hlobil on 19 July 1996, Geneva, and Supertran on 16 July 1996, Geneva.

[47] Questionnaire from Kinrade (1996).

[48] Interview with Uezono, 19 July 1996, Geneva.

may be able to nurture positive relations with Environment Departments, the overall influence of this department compared with Trade and Industry Ministries, for example, may be limited. NGO imprints upon early policy formulated by Environment Ministries can be diluted or countered by government departments representing industrial interests, once they enter into the debate.

In many states, separate Environment Ministries do not exist and environmentalists are 'still kept well away from decision-making' (*ECO*, issue 5, Chantilly 1991). Even where new departments are in existence, they are not conferred authority and political clout (McCormick 1989), and whilst public concern about the environment may be strong, there are few mechanisms for channelling it into the system (Fisher 1995). Moreover, many environmental NGOs either have not mobilised at all on the climate issue[49] or have adopted a strategy that places wider pollution concerns above climate change.[50] The main regional institution supporting Central and East European NGOs, the Environment Centre in Budapest, does not provide funds for campaigns dealing with global environmental problems, offering a strong disincentive to engage with the climate issue (*ECO*, issue 1, New York 1993). One of the key purposes of the CAN network has therefore been to encourage NGOs to be more active on the climate change issue.

One response of environmental NGOs to these limits has been to forge closer alliances with more influential groups in the policy process. As well as more conservative groups such as EDF and NRDC collaborating with business on climate issues to further their own aims (Eikeland 1993c), one channel that Greenpeace has been pursuing is encouraging financial investors to divert investment capital away from fossil fuel energy and into renewable energy production systems (Baird 1996; Leggett 1995a, 1995b). By focusing on the insurance industry, which is especially vulnerable to increased claims from climate-related damage, Greenpeace has attempted to mobilise influential strategic partners in its campaign for precautionary action on climate change; actors who are more likely to receive a sympathetic ear from government than are environmental activists, and whose choice of investment matters a great deal to states. As *Greenpeace Business* (1993a:4) notes, 'the government is fully aware that the London insurance world is a major employer and contributes handsomely to the UK's invisible earnings'. Greenpeace has also been lobbying for the financial industry to organise a presence at the international climate talks in order to have their interests represented, and this occurred for the first time at COP1 (the first Conference of the Parties to the Convention) in Berlin in March 1995.

Greenpeace has also been able to foster an alliance with a number of influential clean energy companies and trade associations, such as The European Association for the Conservation of Energy (EUROACE) and the European Wind Energy Association, in calling for more action on climate change (*Greenpeace Business* 1993b). Climate Network Europe has close relations with COGEN, the corporate umbrella group promoting the interests of the cogeneration energy industries, and

---

[49] Interviews with Kranjc, 15 July 1996, Geneva, and Nyirabu, 17 July 1996, Geneva.
[50] Interview with Hlobil, 19 July 1996, Geneva.

has supported their efforts to promote this form of energy at EU level and in the international climate negotiations.[51] Alliances with corporate lobbies open up new avenues of influence upon government policy and by tilting the balance towards those in favour of action, they are considered by some NGOs as their most effective channel of influence (Leggett 1996).

6.6   Negotiation-bargaining

Lobbying at international conferences did not become a major priority for NGOs until the mid to late 1980s, by which line it was recognised that action to address global environmental problems would increasingly be orchestrated at the international level (Grubb et al. 1993; Princen 1994). This section looks for evidence of 'clear examples where NGO pressure and authoritative contributions can be linked to shifts in delegates' views and policies' (Rahman and Roncerel 1994:241), but also attempts to explore less visible influences, in keeping with the approach employed throughout the book.

The process leading up to the signing of the Framework Convention on Climate Change in June 1992 was intended to involve the active participation of a broad spectrum of NGOs on an unprecedented scale.[52] There was to be no negotiating role for NGOs in the work of the Prepcom,[53] but NGOs were entitled, at their own expense, to make written presentations in the Prepcom process and, conditional on the will of the chair, address plenary meetings of the Prepcoms (Chatterjee and Finger 1994:84). It is noteworthy that although access was permitted to most of the main negotiating arenas, special meetings were sometimes called ('informal informals'), where key decisions were made and from which NGOs were excluded (Andresen 1992; Arts and Rüdig 1995; Susskind 1994). In instances such as these, 'NGO representatives could only stalk the entrance of committee rooms throughout the night in hope of making sure no lobbying opportunity was lost, or, rather more likely, catching a word with some weary diplomats leaving the room in the early hours to gather some information on what was going on' (Arts and Rüdig 1995:484). They could often rely on NGO representatives on national delegations to leak information to them about the progress of the meeting, but there was little they could do during the meeting because of the lack of direct access to officials.

---

[51] GermanWatch in particular enjoy very close relations with the European Business Council for a Sustainable Energy Future (EBCSEF 1996).

[52] Their participation was requested by General Assembly Resolution 44/228 of December 1989, which called on 'relevant NGOs in consultative status with the Economic and Social Council to contribute to the conference as appropriate' (quoted in Chatterjee and Finger 1994:81). Access was not restricted to NGOs with consultative status with the UN ECOSOC (Economic and Social Council) in relation to UNCED; all groups were granted the same rights. For more on NGOs status and roles within the UN see Willetts (1993, 1995) and Domini (1996).

[53] Conca (1995) notes, however, that Greenpeace, EDF and ENDA were able to get individuals onto the working parties of the Prepcom sessions.

Nilsson and Pitt (1994:73) argue that NGOs were unable to 'effect much control over the pace, timing and contents of the INC agenda', because 'Much of the real action took place in ancillary private meetings to which NGOs could not yet gain access, or more informally in the closed circle of decision-makers which did not include the NGOs'. The second meeting of the Conference of Parties to the Convention in Geneva represented a continuation of this process, as the main outcome of the conference (a ministerial declaration) was negotiated in closed-door sessions from which NGOs were excluded (Newell and Paterson 1996). The final deliberations on the Kyoto Protocol appear to have been concluded in much the same manner (Newell 1998).

Nevertheless the climate negotiations in particular were regarded as being at the forefront of attempts to open up international negotiations to NGO participation,[54] in that all sessions (with the exception of the 'informal informals' mentioned above), including working groups, were open and several countries included NGO representatives in their delegations (Dowdeswell and Kinley 1994:120).[55] WWF-Switzerland even had direct access to the Open-Ended Working Group, which drafted the Ministerial Declaration of the Second World Climate Conference, as well as being part of the Swiss national delegation during the scientific and ministerial components of the conference (Boyer 1997). Conca (1993:450) considers that sitting on negotiating delegations is the most important vehicle for NGO influence, and on this basis argues that NGOs on the delegations of more powerful states such as the US have enjoyed the greatest influence over policy at this level.

For Princen (1994:37), 'In negotiations with few precedents, little predetermined structure, an ill-defined agenda and fuzzy outcome expectations [such as those on global warming] simply sitting at the table confers influence.' Yet although a case can be made that being a member of a delegation provides NGOs with a position from which to exert influence, it is also possible to argue that states have relatively cohesive political goals that are not amenable to significant modification through international bargaining, and hence the degree of influence that NGOs can exercise remains restricted. There is a perception among NGOs that states determine much of the scope for compromise and negotiating space before international meetings

---

[54] According to Dubash and Oppenheimer (1992:272) the ability of NGOs to influence international negotiations is growing because of the precedent that was established by their involvement in the UNCED Prepcom meetings. Three-quarters of the NGOs asked in the survey quoted in Chatterjee and Finger (1994:96), however, said they were not satisfied with the access granted to them, and only 12 percent described themselves as content with the amount of speaking time that they were allotted. An indication of the discontent felt by many NGOs about their involvement in the INC process is provided by a comment in the NGO bulletin *ECO* (issue 1, April 1993), which describes 'NGOs huddled around the coffee bar down the hall which seems to be one of the few places we're still allowed to go'.

[55] Canada, for example, included several representatives of environmental NGOs, including the Sierra Club's Louise Comeau, in its COP2 delegation, and the Finnish delegation at COP2 included a representative of Friends of the Earth (interview with Nurmi, 15 July 1996, Geneva).

take place, and that there is little that pressure groups can do to further their agenda once the negotiations have begun.[56]

By the time of the Second World Climate Conference (SWCC) in 1990, environmental NGOs were already suggesting their role in the negotiations was 'minimal' (*ECO*, issue 1, Geneva 1990).[57] There may have been some basis to this. The fact, for instance, that by 1989 the US delegates had already been instructed about their objectives for the climate negotiations was revealed by a leaked 'Talking Points' memo to the US negotiators, outlining which issues to focus attention upon and which to deflect attention away from. The goal of a weak Framework Convention was articulated at the outset of the negotiations (*ECO*, issue 7, Geneva 1990; Hatch 1993). For Lunde (1990:29), 'Government officials travel to each international venue with mandates that are almost 100% fixed' and their agendas are 'even harder to change once final negotiations have started.' It is also argued that crucial decisions, such as those relating to Article 4.2 of the convention or to the final texts of important conferences, are made away from the official negotiating fora, where 'you can see the control national capitals have'.[58] Sieghart is more positive about states' room to manoeuvre, however: although 'many decisions are quite clearly being made elsewhere' and delegations have 'bottom lines', there remains 'scope for negotiation'.[59]

Given the global nature of the issue, many key decisions still have to be made at international meetings[60] (even if mainly between ministers), and hence there are opportunities for corridor lobbying[61] and to maintain pressure on sympathetic delegations to hold their position.[62] A more basic point is that the negotiations bring together all the key players in the same place at the same time, and therefore opportunities are available for direct access to policy-makers, which is often difficult for many NGOs at the national level.

Yet analysis of NGO influence at this stage in the development of a regime, needs to be prefaced by acknowledgement that only a certain number of groups campaigning on global warming are involved in lobbying at this level, and that less organised or powerful NGOs are 'screened out' (Finger 1993:46). For example, while all major international NGOs were officially present as observers at the plenary sessions of the INC, many Southern-based groups, because of limited access to telecommunications, travel and lack of a detailed understanding of the operation of the UN's

---

[56] Interviews with Lashof, 18 July 1996, Geneva, and Singh, 17 July 1996, Geneva.

[57] It is also clear that NGOs feel that the agenda is in many ways determined by powerful industry interests, which 'for the most part ... can get what they want in private months before the bell rings to call delegates into the negotiating sessions' (*ECO*, issue 5, Geneva 1991).

[58] Interview with Supertran, 16 July 1996, Geneva.

[59] Interview with Sieghart, 17 July 1996, Geneva.

[60] Interview with Cozijnsen, 15 July 1996, Geneva.

[61] Interview with Colat, 16 July 1996, Geneva.

[62] Interview with Lashof, 18 July 1996, and Singh, 17 July 1996, Geneva.

bureaucratic machinery, were excluded from the process.[63] On the whole, Northern NGOs have also enjoyed more privileged access to their governments and intergovernmental organisations than is the case with their Southern counterparts.[64] The net effect of the UNCED system, which encourages the formulation of common positions by groups of interests, has been to reduce very disparate demands to the status of a lowest-common-denominator set of diluted policy suggestions. This makes it easier for governments to reject the proposals out of hand, on the basis that they are not directly relevant to their national situations (given that they reflect an international position), or respond to them via tokenistic, incremental policy changes as opposed to the more substantive changes that many NGOs are pursuing. The amalgamation of NGO positions favours the more predominant Western interpretations of the issue, given the greater presence of Western groups at international meetings. This creates problems for Southern groups, which are accused by their own governments of selling out to a Western agenda on the issue.[65]

Moreover, despite the existence of a goal to which nearly all the environmental pressure groups campaigning on this issue could subscribe – that the negotiations should produce commitments to reduce emission – many conflicts were subsumed beneath this embracing goal (Singh 1994). For example, a divergence of perspective between NGOs from North and South became apparent at INC1 (Chantilly, February 1991), where US NGOs came with a position paper that called on the OECD nations to reduce energy-related CO_2 emissions by at least 20 percent by the year 2000, and on LDCs to limit the increase in their aggregate CO_2 emissions to 50 percent over the 1988 levels by the year 2000. A number of Southern NGOs complained that they had not been consulted about the position paper prior to the conference, fearful that aligning themselves with this position would cause their own governments to see them as tools of Northern NGOs (Porter and Brown 1991).[66]

There is a need also to be sensitive to differences of influence among NGOs from the North and South. It can be argued that international groups such as Greenpeace and WWF exert more influence at the negotiation-bargaining stage of the policy process because of their operating reach, which allows them to lobby a number of delegations simultaneously and put pressure on national capitals that are blocking

---

[63] It is also notable that the UNCED model of participation is based on lobbying rather than consultation, a situation that favours the more organised and better resourced NGOs with experienced lobbyists (Carr and Mpende 1996; Finger 1994).

[64] Groups from the North also have institutionalised inter-group contacts and a longer history of campaigning on global issues (Rahman and Roncerel 1994).

[65] This accusation is leant further weight by the funding of Southern NGOs by Northern governments and funding agencies (Singh 1994). For more on the political difficulties involved in North–South collaboration between environmental NGOs see Eccleston (1996).

[66] Jakobsen (1997) shows how connections with the international environmental movement have created problems for Brazilian NGOs working on climate change, because of government suspicion about external interference.

progress.[67] This can be contrasted with national-based groups, where the effectiveness of lobbying is mainly restricted to their home nation.

Chatterjee and Finger (1994) further distinguish between Northern political ecology groups such as Friends of the Earth and Southern participatory coalitions on the one hand, and better resourced Southern groups such as Third World Network and Centre for Science and Environment (CSE)[68] and the 'Big 10' (Mitchell 1991; Wapner 1996) US environmental groups on the other. The former are regarded as being more marginalised in the international process, and the latter having significantly more access. In this way, the North–South dichotomy might be more usefully broken down into an 'insider/outsider' distinction that cuts across geographical boundaries.

Using a different metaphor, Conca (1995) suggests a concentric circles model for thinking about NGOs' access and power within the UN system, with groups such as Greenpeace and WWF at the centre, and NGOs from LDCs towards the fringe. The 'insider' groups are able to be present at the meetings where much of the politicking takes place. As Chatterjee and Finger (1994:96) observe, 'having some influence on the negotiations was basically a matter of good relations with government delegates and the secretariat', the sort of relations that are enjoyed by groups such as WWF, International Union for the Conservation of Nature, World Resources Institute and the 'Big 10', with their 'substantial lobbying power and access to the negotiations' (ibid.:70).

For example, EDF's senior scientist chaired the Advisory Group on Greenhouse Gases (a working group on limitation strategies for greenhouse gases), and its primary international lawyer represents EDF on two advisory bodies on the atmosphere and oceans to the secretary-general of UNCED. In 1990 EDF attorneys also participated as observers at the IPCC meetings at the SWCC, actively contributing both to the ministerial declaration and to the scientific and technical statements at preparatory meetings of the INC. These observations lend weight to Raustiala's (1996:56) claim that, of the NGOs campaigning on climate change, 'many US based NGOs, because of their size, expertise and influence on the government of the US were particularly influential'. A similar pattern emerges in the South, where 'insider' groups such as CSE and BCAS are able to ensure that interventions from Indian and Bangladeshi ministers, respectively, are often taken verbatim from their research.[69] BCAS claimed to have actually drafted some of the interventions from the Bangladeshi delegation at the COP4 meeting in Buenos Aires.

A powerful counterpoint to the 'insider/outsider' or 'concentric circle' formulation of influence is the case of Global Commons Institute (GCI).[70] GCI takes as given that economic growth is driving the global community 'over the thresholds of global ecological stability' (GCI 1994:7), and that there is a need to move

---

[67] Interview with Lashof, 18 July 1996, Geneva.
[68] Questionnaire from Sharma (1996).
[69] Interview with Rahman, 19 July 1996, Geneva.
[70] The GCI was founded in 1990 after the SWCC, and has taken an active part in the IPCC and INC process.

immediately towards a minimum 60 percent cut in global CO_2 emissions. However it admits that 'The structural and restructuring implications of this are great' (ibid.:2). GCI first became involved in the climate debate by challenging the use by IPCC Working Group 3 of cost–benefit analysis as the means by which to assess the worth of different policy options. In this sense GCI was directly confronting a 'given' of environmental policy-making: the use of cost–benefit analysis for the economic costing of policy options. It could therefore be expected that it would find success hard to achieve.

However Peter Sturm, head of the OECD's resource allocation division, said that, 'GCI should be very pleased with the influence they had' in the IPCC WG3 debate on cost–benefit analysis (Douthwaite 1994). Critiques provided by GCI resulted in many LDCs calling for the withdrawal of the chapter by David Pearce on the economics of climate change (see below). GCI was therefore successful in challenging the prevailing assumptions in the policy debate about how to value action on climate change, nurturing a more normatively grounded discursive terrain subversive of the previously assumed legitimacy of cost–benefit analysis.

GCI has also successfully achieved some success in promoting its 'contraction and convergence' approach whereby the highest emitters have to contract their emissions of CO_2 and the lowest emitters are permitted to increase their emissions, converging on an optimal level of carbon usage that would prevent dangerous interference with the climate system (the stated goal of the convention). The idea has been carried forward by a number of LDCs[71] as well as attracting the support of Northern sponsors (Corner House 1997). The parliamentary group GLOBE (Global Legislators for a Balanced Environment) adopted the idea and actively promoted it in the climate negotiations, providing a key platform for GCI in the policy debate. The alliance between GCI head Aubrey Meyer and Tom Spencer MEP, head of the GLOBE grouping in the EU, in particular, has provided GCI access both to key decision-makers and the resources and formal respectability that are associated with an accepted and established group such as GLOBE.

The influence that such a small and poorly resourced group can have upon policy, suggests that access to and reserves of resources do not, of themselves, confer influence. So whilst the system that operates at the UN is likened by Chatterjee and Finger (1994:70) to the US model of participation in the policy process, where access to capital and a professional modus operandi are prerequisites for influence, the GCI example does seem to illustrate that even when these conditions are not satisfied, influence can be exerted.

The example also suggests that making an impact at this stage of the policy process is partly a question of being able to provide policy-makers with relevant and strategically useful information and advice. In this respect, Rahman and Roncerel (1994:241) note that, particularly in the early rounds of negotiations, NGOs' knowledge of the science and politics of climate change enabled them to make 'notable

---

[71] The African group of nations have taken up the proposal as has the Chinese government. See Statement of African Group of Nations, August 1997, and Sung (1997).

contributions', bringing expertise to bear that would otherwise have been unavailable to states. The lack of informed critical assessment of the various policy options being discussed at the negotiations led to another alliance, this time between environmental NGOs and the G77 and China on the issue of joint implementation (JI). The G77 and China initially supported the idea, but went on to reject it at COP1 when alerted by NGOs about the negative way in which JI schemes could affect their interests. Arts (1998:142) quotes a government official from this bloc: 'Environmental organisations very clearly helped us to understand that JI was looked at by the industrialised countries as a means to avoid reductions in their own countries.' These parties were able to draw on arguments made by environmental NGOs to insist that credits should not be awarded for JI schemes; an intervention that ensured that JI schemes were accorded pilot status only. Grace Akumu of Climate Network Africa (CNA), argues that the position of many African states at COP1 against accreditation of the JI projects of Annex 1 countries was attributable to CNA's warnings that Annex 1 party commitments had to be strengthened before acceptance of the proposal.[72] The fact that governments' own positions were both ill-informed and underdeveloped on the issue, made them amenable to the advice and pressure of environmental NGOs, even at this stage of the policy process.[73]

Similarly, the CSE report 'Global Warming in an Unequal World' published in 1990, performed an agenda-setting function in the international negotiations by sparking a debate on per capita emission entitlements. As Rahman and Roncerel (1994:250–1) suggest, the point is not that issues 'would not have come into the negotiations at a later time or in some other way without the NGO debate', but that groups such as CSE were able to raise the profile of this debate earlier in the negotiations than would otherwise have been the case, and affected its subsequent course.

Climate Action Network South Asia (CANSA) also drafts issue-based or regional position papers and country papers with a view to strengthening the ability of government negotiators from the south Asian region, to negotiate more effectively. In this situation, as with the work of CNA in Africa, NGOs are generally very supportive of their governments' positions; a fact that generally makes governments more receptive to the advice and research of these groups. Akumu describes CNA's work with their governments as a 'partnership'.[74] As Banuri (1993:64) notes, the net effect of NGOs forming alliances with these governments may be to tilt 'the balance ever so slightly towards the normally weak Southern delegations'.[75]

These examples seem to illustrate the point that as groups acquire greater technical capability, at least in the eyes of governments, they are accorded greater access,

---

[72] Interview with Akumu, 17 July 1996, Geneva.
[73] This was perhaps especially so for many LDCs because industry lobby groups had not yet mobilised to counter NGOs' efforts (see Chapter 5).
[74] Interview with Akumu, 17 July 1996, Geneva.
[75] This resonates with Nye and Keohane's (1973:xvii) argument that transnational relations have the effect of increasing the ability of certain governments to influence others.

derived from their ability to provide technical research input of immediate and obvious use to the negotiators. The provision of studies on the regional impacts of climate change and other technical and policy-relevant research by CAN has undoubtedly aided its integration into the decision-making process at the international level.[76] There is support for this view from government officials. For instance Philip Dykins of the UK Department of the Environment Global Atmosphere Division argues that 'CAN produce a lot of very good stuff, quality documents which we use. These people are obviously going to have more influence than Greenpeace scaling the walls outside.'[77] In this regard, Dubash and Oppenheimer (1992:278) find that CAN has 'proved a constructive model for interaction between governments and NGOs in finding creative yet immediate solutions to global warming'. Hence, though it is by no means always the case, as the example of GCI clearly illustrates, it is notable that many of the groups that are in a position to support their arguments with substantial technical expertise and research, are often 'insider' groups such as BCAS and EDF. Besides providing support for their own country's position in the negotiations, the deployment of scientific and technical advice to friendly states can also be used to support NGOs' preferred negotiating positions. Identifying states that will potentially serve as collaborative partners, however small or seemingly peripheral to the negotiations, is a way of greatly influencing the debate, given that every state has an automatic right of access to committees and working groups from which NGOs are excluded.

In the context of the climate negotiations, this has taken the form of groups such as Greenpeace nurturing a special relationship (Rahman and Roncerel 1994:245) with AOSIS (the Alliance of Small Island States): the group of states most vulnerable to climate change. Greenpeace and FIELD (the Foundation for International Environmental Law and Development) have been able to provide scientific back-up as well as policy and strategic advice to strengthen AOSIS's pro-action negotiating stance. As Philip Dykins, UK delegate at the climate negotiations, notes, 'Some of the small island states have limited resources and invariably rely on Greenpeace International to help them formulate some of their arguments.'[78] From the SWCC in 1990, 'Small island states were looking for genuine partners to voice their concerns, [they] discovered that the NGOs participating in the climate negotiations could be constructive partners and conduits of their concerns' (Rahman and Roncerel 1994:243). Sefania Nawarda from the Fijian Ministry of Local Government and the Environment notes in this respect, 'The government department I work for is small, and a lot of the work we do would have to be done by NGOs, so there is quite close cooperation with them.'[79] He continues, 'we are closely involved with NGOs like FIELD, Greenpeace and WWF ... our major difficulty

---

[76] Interview with Capoor, 15 July 1996, Geneva. See for example CNA's (1993) 'Study of the IPCC greenhouse gas inventory methodology applied to land use and forestry changes in Kenya'.
[77] Interview with Dykins, 25 February 1994, London.
[78] Interview with Dykins, 25 February 1994, London.
[79] Interview with Nawada, 16 July 1996, Geneva.

is our isolation from the major centres of research and information, so this is a major role we see NGOs playing'.[80]

More directly still, the AOSIS protocol (based on the Toronto target), which was suggested as an appropriate policy instrument to complement the convention, is widely acknowledged to have been drafted by the environmental law group FIELD, despite the reluctance of governments to admit that they permitted an NGO to help them draft the proposal (Arts 1998). FIELD modestly claims that 'The AOSIS Protocol was drafted with assistance from [us] to reflect the most politically feasible and environmentally progressive approach to strengthening the Convention' (quoted in ibid.:136). The drafting of the protocol served to inject a greater sense of urgency into the debates taking place at INC11 and COP1 about a legally binding instrument post-Rio (ibid.).

The ability to provide strategically useful information helps to explain how GCI has been able to form alliances with states such as India and Chile, and enjoys the 'widespread support of non-OECD nations' (Douthwaite 1994). For GCI head Aubrey Meyer, the partnership is strengthened by the fact that many non-OECD countries are 'mostly deprived of impartially [sic] analysed information about this matter' (Meyer 1994). GCI was able to convince IPCC WG3 to remove a key chapter submitted to it by Professor David Pearce, because 'a number of third world countries including India, complained after being alerted by Aubrey Meyer' at the Montreal 1995 meeting of the group (Meyer and Cooper 1995).

Subsequently, the summary for policy-makers that the IPCC adopted in Montreal 'did not use the economists' figures because of Meyer's lobbying' (ibid.; Masood and Ochert 1995). For Masood and Ochert (1995:225) the connection is clear: 'GCI persuaded those responsible for the summary for policy-makers to erase references to damage estimates and include phrases such as "the literature on the subject is controversial", mention of "the value of life" and reference to the fact that the "loss of unique cultures cannot be quantified".' At the Geneva meeting of the IPCC prior to this, GCI had circulated corrections to the draft report to which Pearce had contributed, based on the 'unequal life-evaluation used in the economist's work' (Meyer and Cooper 1995). Cuba and Brazil took up GCI's criticisms and objected to the inclusion of Pearce's work, forcing the arrangement of a meeting in Montreal in October 1995. GCI's analysis of the climate change problem was also 'taken up by the Indian environment minister on behalf of many developing nations' (Jaeger and O'Riordan 1996:6).

As a collective entity, CAN was also successful at COP1 in helping to mobilise the support of China and India for a protocol to the convention, by bringing them into the 'green' group it had been active in crafting (Arts and Rüdig 1995). As Grubb (1995a:4) notes, 'NGOs probably played a significant role in persuading India and Brazilian delegates to make the moves that broke the impasse and that led ultimately to the developing countries' "green paper" and thence to the [Berlin] mandate.' Influence was brought to bear through media exposure, the tabling of papers and,

---

[80] Ibid.

most notably, the mediating role that some NGOs were able to perform between factions within the broad North–South groupings (Arts 1998). Rather than direct impact on the closed door diplomacy that gave rise to the Berlin mandate, influence was exercised more subtly by creating a negotiating space in which agreement on a mandate was possible.

Indeed, looking just at the officially recognised and 'visible' roles performed by NGOs only tells us so much about the scope of their overall influence. Many NGO interactions with policy-makers take the form of corridor lobbying, behind the scenes selling of diplomatic packages and use of personal contacts. This section will therefore attempt, where possible, to include analysis of the exercise of less official channels of influence upon the international negotiations, as well as develop the points made at the start of the chapter about environmental groups, in order to offer a broader perspective on the influence of positional factors.

One of the key objectives of environmental pressure groups at international meetings is to improve their transparency, based on the assumption that greater accountability to domestic constituencies will increase the pressure on politicians to act. By dispersing to national and local audiences news of what governments have and have not managed to agree in international fora, momentum for change generated domestically can be harnessed to pressure for action at the international level. This is the thinking behind the creation of the publication *ECO*, which CAN produces at all international meetings and disseminates to a wide audience (Arts and Rüdig 1995).[81]

A similar recognition of the need to draw attention to proceedings in international fora in order to activate domestic opinion, accounts for the stunts that groups perform at conferences to generate media publicity on the discussions and issues being debated. One example is the Greenpeace blockade of the main entrance to the UN with a lorry carrying a banner pronouncing 'Bush blocks climate treaty, Danger: global warming ahead'. In addition six activists chained themselves to the top of flagpoles near the conference centre, unfurling another banner that read 'Bush: Heat up the Economy, Not the Planet – Cut CO_2 Now!' (*ECO*, issue 3, New York 1992).

NGOs also attempt to press upon delegates at international meetings the expectations of the public, warning them of the public relations fiasco that might ensue in the absence of concrete action. These sorts of pressure are visibly applied during the international climate negotiations, though their effects are difficult to ascertain. NGOs warned delegates at Nairobi in 1991 that 'The entire world is awaiting a successful result to these negotiations' (*ECO*, issue 10, Nairobi 1991). Tessa Robertson (Greenpeace-UK), then of WWF-UK, asked delegates at the spring 1992 meeting in New York, 'At the end of this, what are you going to tell your publics what this Convention does for the environment?' (*ECO*, issue 5, New York 1992). In

---

[81] Sieghart argues that '*ECO* has quite considerable influence in that [delegations] all read it' (interview, 17 July 1996, Geneva). Because of this, delegations are sensitive about the way in which their positions are portrayed in the newsletter. Norway, for example, challenged *ECO*'s representation of its positive endorsement of JI as 'running for cover' (Betsill 1998).

this way, groups are able to construct a yardstick against which the outcomes of negotiations will be assessed, as well as nurture an expectation on the part of the public that those demands will be met, through use of the media.

One other channel of less visible or empirically demonstrable influence that environmental NGOs benefit from, relates to the reactions that governments anticipate from NGOs and the public if they decide to pursue, or fail to pursue, particular courses of action. Based on their claim to represent broad constituencies of public support, environmental NGOs attempt to demonstrate to governments that courses of action that widely deviate from what is expected will be unpopular, thereby providing an incentive for states to take action. Susskind (1994:127) argues that 'This can help even the most powerful leader anticipate national and international reactions and gauge the acceptability of various negotiating postures more effectively before public pronouncements are made.' Rahman and Roncerel (1994:241) observe that delegates at the climate negotiations 'found it quite useful to discuss and test out some of their initial ideas with NGOs ... the most controversial issues were not even brought to the main forum of the negotiations after consultations between the delegates and the NGOs'. After environmentalists attacked the concept of a 'pledge and review' as a 'hedge and retreat' strategy, many governments withdrew support for the proposal. The incident provides an example of governments factoring into their policy thinking anticipation of the reaction that may greet a particular form of action. It is thought that US NGOs in particular 'may have contributed to its demise after only a short life in the negotiations' (ibid.:264).

One way of trying to assess the degree of influence that NGOs have had over policy at the international level, is to look at how NGOs themselves perceive the situation, given that they are in the business of employing their (often scarce) resources in the most efficient manner and therefore target areas where they think they may be successful. In order of priority, meeting with delegations and submitting written evidence feature highly in what NGOs perceive to be a hierarchy of influence. Very few of the NGOs that participated in a survey on the UNCED process (quoted in Chatterjee and Finger, 1994:96) thought that their views had been represented in the legal documents produced. Finger (1994) argues that NGO reports and advice inevitably feed into governments' existing frames of interpretation. He notes, that 'If NGO wording was adopted at all, it was incorporated into the texts in an almost token manner and in disregard of NGOs' underlying arguments' (ibid.:208).

Indeed many of the NGOs' key concerns left no imprint on the Climate Convention. As Rahman and Roncerel (1994:263) argue, 'The bitter pill of an absolute reduction in consumption in order to reduce greenhouse gas emissions to the atmosphere was never explicitly prescribed in the Convention.' Even minimal goals, such as the achievement of a *binding* target for industrialised countries to stabilise their CO_2 emissions at the 1990 levels by the year 2000, remained unfulfilled. Arts (1998) credits NGOs with some indirect political influence over Article 4.2(a) and (b) on the basis that they helped to ensure that the convention was not weaker than it would have been if they had not been active alongside the EU, AOSIS and the G77. Despite the NGOs' failure to elicit the meaningful reductions that they were hoping for, however, the concept of ecological limits, which they wanted to underpin the purpose of the convention, was incorporated into the stated objective of the convention.[82] Though this could equally be argued to be the result of the work of the IPCC Working Group 1 in defining a suitable goal for the convention, Arts (1996:20) finds that NGOs and government representatives believe that environmental groups contributed to the development of this objective.

Lunde (1990:25) also argues that the inclusion of the precautionary principle in the preamble of the convention was an 'important victory' for environmental NGOs. As with the above point, it could be argued, as Litfin (1994, 1995) does, that this could be the result of the work of the IPCC and cannot be clearly linked to the lobbying of environmental NGOs. However environmental NGOs might be able legitimately to claim that they prepared the ground for the acceptance of such a principle as a norm to guide policy. Similarly the NGOs' emphasis on the responsibility of the North is enshrined in the convention in a number of places, but it is difficult to know whether this can be attributed to the influence of NGOs, given the obvious discrepancy between the North and South in terms of contributions to the problem and the centrality of this principle to the policy discourse of many developing countries. Arts (1998:121) suggests that 'Although some NGOs such as CSE and Greenpeace were able to contribute to the debates on principles, their political influence on Article 3 was very limited.'

It is of course impossible to assess accurately the extent to which these features of the Climate Change Convention were a response to their continually being pressed upon delegates by NGOs, or whether other factors were more important in explaining these formulations, that is, whether they would have been included anyway. In situations where there is a narrow margin in favour of a particular option, NGOs feel they can make a difference. Arts and Rüdig (1995:486) note that while 'It is difficult if not impossible to quantify the influence of environmental NGOs at the Berlin summit, the Berlin summit did produce a mandate, and there were no doubts in their minds that it had been a close-run thing in which they had made a difference.' It is unrealistic to expect a precise, scientific assessment of their influence however. A quote from Mark Valentine seems to sum up the situation neatly with regard to observable NGO influence on the convention process: 'Bits and pieces were tinkered with and modified here and there, but the structure of the agreements, the context within which they were considered, and the level of political and financial investment all conformed to governments' expectations not NGOs' (quoted in Chatterjee and Finger 1994:96).

NGOs were nevertheless successful in creating a public expectation that a convention would be ready for signature at Rio in June 1992. It is clear that governments had at the forefront of their minds, the danger of a public relations fiasco at the

---

[82] The ultimate objective of the treaty, stated in Article 2, is to 'prevent dangerous anthropogenic interference with the climate' system' (UNFCCC 1992:2).

Earth Summit. Brenton (1994:191) notes that in the US, for example, 'President Bush's staff were acutely conscious of the political price (in an election year) of wrecking Rio.' It is clear, moreover, that NGOs 'helped to maintain the momentum of the INC', 'encouraged the cultivation of a consensus' around particular policy goals and succeeded in 'setting the tone and maintained pressure towards a meaningful and universally acceptable Convention' (Rahman and Roncerel 1994:255, 257). The distinction that Arts (1996) draws between the impact of NGOs on the *process* of forming the convention on the one hand, and their lack of direct influence on the actual text on the other, supports the position here that much of the influence of environmental groups at the international level has not been captured in the convention itself. That most influence occurs during the shaping and formulation of policy, rather than being visible in final policy outcomes, should not be surprising, as it remains the principal task of governments to make the final decisions. It is also not a valid basis to dismiss the politics that precede the final, formal stages of negotiation. Outcomes are the product of contestation among different parties and bear the imprint of NGO activity, even if NGOs are not in the room when the final text is agreed.

The problem of assessing influence is complicated by that fact that many of the most influential groups are also those closest to governments, those whose policy prescriptions are more easily accommodated within existing frameworks for understanding and responding to the threat of climate change. These groups' positions may be 'so close to the positions of governments that their distinctive impact can hardly be detected in the texts' (Chatterjee and Finger 1994:98). Making a similar point, Athanasiou (1991:13) argues that the larger, more influential groups derive their bargaining position on the global warming debate from an ability to 'avoid stickier political issues'. Given the areas of overlap between certain NGO and government agendas, it is more difficult to assess which policies can be explained in terms of NGO influence and which, during their initiation or subsequent development, drew nothing from NGO pressure but were incremental changes the government would have made anyway.

The clearest example is that of the Maldives government, which gives the climate change issue the highest priority, complicating assessment of the degree of influence exerted by groups such as Greenpeace, even when government officials acknowledge their input.[83] The dilemma boils down to a counterfactual argument, not subject to empirical verification, about whether a particular piece of text or a preferred interpretation would have emerged in the absence of influence. For the purposes of this analysis, it is sufficient to say that if policy-makers who are the target of influence acknowledge that their positions have been either moved or consolidated by the lobbying of environmental NGOs, then it can be established that groups have made a difference, the principal disclaimer being that the government in question was receptive to that influence and therefore did not have to be pushed very far.

---

[83] Interview with Khaleel, 17 July 1996, Geneva.

6.7 Implementation

It is widely recognised that NGOs are already playing an important part in helping to enforce the Climate Convention (Grubb and Steen 1991; Nilsson and Pitt 1994; Sands 1993), and will form an integral part of the enforcement system that will ultimately enforce the Kyoto Protocol. Implementation of the convention is, for the most part, demonstrated via national reports on action being taken to reduce levels of greenhouse gas emissions, which are reviewed internationally for their adequacy. This takes the form of secretariat officials visiting countries that are party to the convention to 'meet with government officials and NGO representatives in order to assess the country's climate change policies' (Anderson 1995:11). Consultation with the secretariat provides NGOs with an opportunity to highlight weaknesses or 'silences' in the national reports, and to comment on whether they consider the time-tables included to be realistic, in light of other government actions. The potential for environmental groups to embarrass governments in this way creates an incentive for the latter to anticipate criticisms, and to endorse a broader range of policies than would otherwise have been the case.

NGOs have been engaging in the review of national action plans by presenting their own independent evaluations of communications submitted under the convention. For example COP2 saw the publication of the fourth review of national action plans on climate change by the CAN network (CNE/CAN 1996). These evaluations have encouraged the direction and focus of the secretariat's assessment of the extent to which countries are complying with their obligations. By providing 'predigested information', NGOs have been able to guide the attention of the reviewers to particular sets of figures, or omissions, in the national communications. Testimony to the influence of the NGO evaluations is found in the fact that they are widely referenced in governments' own policy documents (Arts 1998).

NGOs have also exposed states' inability to fulfil their commitments under the convention. Greenpeace International's report, 'The EC's next global warming factories', published in April 1994 (*Acid News* 1994b), showed how EU proposals to build new power plants could prevent the EU from meeting its goal of returning CO_2 emissions to the 1990 levels by the year 2000. This sort of exposure provides insights into the wider context of governments' ability to meet their commitments, which governments are often loathe to highlight in international evaluations. As Stairs and Taylor (1992:117) note, 'the only effective forces for compliance at present are reasoned argument and the embarrassment factor'. With regard to the latter, NGOs can play a 'vital role' (ibid.), for as Litfin (1993:109) notes, 'countries with influential environmental movements may have an additional incentive for compliance'. For Arts (1998) this implies process influence rather than political influence on the basis that NGOs' main contribution to the reporting process is to provide depth and transparency to the process, which is difficult to detect in final policy outcomes.

Because the commitments in the convention are couched in such vague terms, interpretations of obligations at the national level are paramount, and here environmental NGOs can once again contribute both expertise and pressure. The norms that will bring the aims of the convention to realisation, can be shaped by the work of NGOs in setting and diffusing precedents for national reporting and policy implementation. Limits are placed on this sort of influence, however, by Article 12.9 of the convention, which gives states the right to keep confidential any information submitted under the convention and the fact that NGOs' formal right to be represented as observers at the sessions of the COP can be revoked if 'at least one third of the parties present object' (UNFCCC 1992: Article 7.6).

One of the ways in which the Climate Convention envisages Southern parties being able to fulfil their commitments, is through the transfer of technologies from Annex 1 parties to non-Annex 1 parties. Here too there is potential for a great deal of NGO influence, by ensuring that technologies are appropriate to local circumstances and not redundant technologies from the North. Overseeing the transfer of appropriate technology will, however, be a difficult task for NGOs. The Global Environmental Facility (GEF), the interim institutional mechanism by which these transfers will be orchestrated, is regarded by some as being inaccessible to the views and representations of environmental NGOs. The Ad Hoc Working Group on Global Warming and Energy, under the Scientific and Technical Advisory Panel of the GEF, provides one point of contact for NGOs to assert their choice of criteria for projects. NGOs may participate in initial project identification through the consultation process that is meant to take place, as well as participate in the technical review panels (Wood 1993:222). Wood notes, however, that 'Beyond these two areas of involvement, the role of NGOs in project development is much less well defined' (ibid.). The dual dangers, based on past experience, that NGOs will only be consulted at the implementation stage of a project, by which time many plans will not be amenable to significant review, and that information about projects will not be made readily available, pose further potential obstacles to influence. The capital-intensive, high-cost nature of projects likely to be funded by the GEF imply a 'strongly centralised implementation mechanism' (ibid.:242) in which governments will adopt most responsibility for project execution.

The 'insider' and 'outsider' distinction employed above in relation to the international negotiations might be applied here, in so far as it is the more radical 'political ecology' groups (Chatterjee and Finger 1994) from both the North and South that are marginalised in terms of access and influence, and the larger groups, such as those belonging to the 'Big 10', that are included in dialogue with financial institutions. Washington-based NGOs such as EDF and NRDC already have regular meetings with the World Bank (*Environment Bulletin* 1995; Princen and Finger 1994:6). The make-up of the GEF steering committee, created in 1992 and composed of NGOs and implementing agencies, would seem to illustrate the point that the larger Northern NGOs are best represented.[84] US policy towards the GEF is also said to be highly influenced by the views of EDF, the Sierra Club and the NRDC, through the monthly meetings that take place between the Treasury Department and these environmental NGOs on GEF-related issues (Gan 1993). The influence of these US groups is likely to be greater than that of others lobbying the GEF because the US provides the largest proportion of funds for the facility, and so key financing decisions will be made by the US Congress and Senate.

The other key institution with which NGOs will want to engage at the implementation stage is the CDM (Clean Development Mechanism). Whilst it is impossible to predict with any accuracy at this stage the nature of the relationship between NGOs and the CDM, it is apparent that NGOs with the appropriate technical expertise will seek to influence the transactions overseen by the institution, as well attempt to expose projects they feel are not worthy of accreditation. Permit trading, once it begins, will also be difficult for NGOs to monitor, particularly if, as envisaged, interfirm trading is permitted. The scale and volume of transactions will require sophisticated networks of surveillance to be in place, if NGOs are to continue to perform watchdog functions effectively and in a way that supplements the capabilities of international institutions. The fact that so many of the trading relations can be undertaken privately potentially removes an element of public scrutiny (through institutional oversight) that has enabled NGOs to influence the implementation stage thus far.

It is notable, in general, that at this stage in the policy process environmental pressure groups find it harder to bring the weight of public pressure to bear on governments, as it is more easily dissipated by the lethargy and complexity of bureaucracy, and by realisation of the scale of costs to be borne by policy options designed to meet international obligations (Grubb 1991). Influencing the implementation of government proposals will prove as difficult as getting backing for the proposals in the first place. Legislatures hostile to international action on climate change can also scupper or water down the provisions that NGOs have successfully lobbied for at the negotiations. Lashof argues, for instance, that in the US 'The limited congressional support for taking stronger action on the issue is probably the biggest single limiting factor to our further influence.'[85] However implementation at the national level also means that environmental NGOs can once again make use of those national level contacts which they feel give them their greatest leverage (Rahman and Roncerel 1994). They can also seek to build domestic coalitions supportive of the necessary legislation to bring the international climate agreements into effect. In the UK, for example, Friends of the Earth (and others) have been promoting (in alliance

---

[84] One way in which the GEF has sought to integrate NGO input is via the NGO small grants programme, where support is provided to NGOs for small-scale activities that address global warming. For more on the official role envisaged for NGOs in the GEF see GEF (1995). Five NGOs are allowed to attend Council meetings and five to observe the meetings. Travel and subsistence is also provided for four NGO representatives from recipient countries (El-Ashry 1995; GEF 1995).

[85] Interview with Lashof, 18 July 1996, Geneva.

with sympathetic industry groups such as the Association for the Conservation of Energy), an energy efficiency bill and an energy conservation bill, and are seeking to push through a road traffic reduction bill, all of which will be essential if the UK government is to meet its 12.5 percent emission reduction target as part of the EU 'bubble' agreed at the Kyoto meeting.

6.8   Conclusions

A great deal of hope is undoubtedly vested in environmental NGOs to activate the required political change to confront global warming. Professor Bert Bolin, chair of the IPCC, disclosed that the experience of dealing with governments in the INC process meant that he now 'looked to the environmental movement, Greenpeace and the others to save us' (quoted in *ECO*, issue 2, Geneva 1990).

It seems from the above analysis of a spectrum of NGO activity, that environmental NGOs can perform different roles at different levels and exert their influence in a variety of ways. There would appear to be little doubt that environmental NGOs, in the politics of climate change, constitute an important force for political change by helping to overcome social inertia and bureaucratic resistance to policy reform. They initiate institutional change and can articulate the norms that they feel should underpin a regime. Through exercise of their bargaining assets, NGOs can secure a place in policy deliberations, though of themselves these assets do not guarantee direct influence over international policy.

It is also clear that, just as the various environmental NGOs working on climate change are positioned differently with regard to their influence with individual governments, the disposition of states to adopt NGO positions also differs. A correlation between 'strong' national environmental lobbies and a proactive negotiating stance on global warming does not appear to be tenable, as the cases of the UK and US illustrate. Success for particular NGOs has to be placed in the context of the difficulty of responding to their demands. Groups with a more far-reaching set of demands cannot be dismissed as unsuccessful on the grounds that fewer of their demands have been acted upon compared with more conservative groups, where the degree of political change sought is significantly less. Small concessions to more politically problematic demands may be harder fought than larger concessions to demands that are more incremental in nature. In other words, groups that appear to force the most concessions from governments are not *necessarily* the most powerful or have gained the most ground in the debate. Groups enter the debate from different starting points, for reasons that relate to the nature of the agenda they are pursuing and the interests of the state they are attempting to influence.

Opportunities to exercise political leverage are also dependent on the issue in question (Bramble and Porter 1992). In the case of global warming, NGOs are confronted with a lack of political will on the part of the states that contribute most to the problem, the presence and active lobbying of some of the most powerful industrial interests in the world economy, a widespread perception of the high economic costs associated with action, and an absence of pressing evidence of the problem to confirm scientific opinion on the issue. These characteristics of the issue disincline it to significant NGO influence. Skolnikoff (1990:895) notes in relation to climate change, that despite the 'growth in mobilisation and influence of national and international environmental movements . . . it remains doubtful that these movements can move publics and national policy processes sufficiently in the light of the uncertainties of the phenomenon itself and the more uncertain costs of action.' Similarly, Philip Dykins of the Global Atmosphere Division of the UK Department of the Environment argues that 'it is difficult to have that great an influence on such difficult political decisions and larger interests. [Environmental NGOs] are in a very difficult position.'[86]

Perhaps the best account of NGO influence that can realistically be advanced is that, 'In the absence of pressure from social movements, nations might never move beyond vague declarations of intent' (Litfin 1993:107). That vague declarations of intent are considered necessary, perhaps also testifies to NGO influence. NGOs can make it more difficult for states to engage in 'routinised non-decision-making' (Breyman 1993:127) to avoid addressing issues altogether. As Fiona Weir, Friends of the Earth climate campaigner, notes, 'what we tangibly achieved is small, but were environmental NGOs not active on an issue, I think even this small amount would not have happened'.[87] Yet the argument here goes beyond such a proposition, by asserting that environmental NGOs can contribute to the formation of perceived interests and force the pace of policy change, and by raising expectations and providing relevant expertise and information they can help to set the boundaries within which decisions are made. In regime terms, they can 'provide different sets of incentives or pay-offs for states' (Nye and Keohane 1973:375). They help to generate the pressures that prompt action and in so doing inform our understanding of regime formation. Their intimate involvement in constructing the provisions of a regime help explain the content and dynamics of change within the regime. And by exposing and reporting non-compliance they contribute to regime effectiveness. Their ability to exploit the connections between domestic and international politics that cajole states into action, is a further undervalued contribution that environmental NGOs make to the day-to-day functioning of the institutional architecture of global politics.

---

[86] Interview with Dykins, 25 February 1994, London.
[87] Questionnaire from Weir (1996).

7

Conclusion: states, NGOs and the future of global climate politics

7.1 Key themes

This book has sought to develop a number of key themes around the question of the political influence of non-governmental actors upon the course of global climate politics. This section will review the main contributions in theoretical and conceptual terms. Section 7.2 will discuss the relevance of each of the 'actor' chapters to understandings of global climate politics, and the ways in which the analysis can usefully complement or challenge existing explanations. Section 7.3 will situate the project in broader theoretical terms within the discipline of international relations. In Sections 7.4 and 7.5, future avenues for research on NGOs in global politics will be suggested.

One feature of this book that other writers in international environmental politics have touched upon, but not comprehensively developed, is use of a *regime breakdown approach* in all but the media chapter (Chapter 4), where it was not thought to be relevant, given that the media are less involved in the negotiating and implementation stages of the policy process. This breakdown, while recognising that the stages identified are not temporally distinct, divides the policy process for analytical purposes into agenda-setting, negotiation-bargaining and implementation. *Agenda-setting* constitutes that stage of the policy process before an issue is dealt with at the international level, when the nature of the problem and the scope of action required are under consideration. The *negotiation-bargaining* stage refers to the politics that take place within formal international fora; the focus is on the process of bargaining and the trade-offs that are traditionally assumed to take place exclusively between states. Finally, the *implementation* stage involves the process of enforceing commitments, mainly at the national level, once measures have been agreed at the international level. This approach, whilst similar to those adopted by Boehmer-Christiansen (1989), Young (1989a), Haas et al. (1993) and Osherenko and Young (1993),[1] suggests that the different stages of the policy process help to illustrate how an actor's influence is heightened or wanes at particular points in the process. Apart from passing reference to this type of formulation by the above writers in the

---

[1] Although these writers use similar categories to those employed here, they are used to describe different parts of the policy process. The writers mentioned here use the term agenda-setting, for example, to denote a process that takes place within international fora. Here, however, it describes a largely national policy process.

literature on global environmental politics, analytical (and empirical) differences in the policy process have not generally been emphasised.

This book has shown that such distinctions, while not always temporally or analytically distinct, do offer explanatory insights. It is argued, for instance, that while the influence of the fossil fuel lobbies, and to a lesser extent environmental NGOs, persists for the whole of the policy process, the influence of IPCC working groups is concentrated in the earlier stages of policy-making. The breakdown also offers a way of addressing the interactions between levels of analysis, between national, international and transnational, and of course the way in which the influence of domestic politics spills over into the international arena, and vice versa (see below).

Particular emphasis is placed on the way in which the particular issues that together constitute the *problematique* of global warming as a political question, fundamentally affect the nature of the political relationships between actors engaging with the issue. This represents a significant departure in writing on international relations and the environment, which, from a regime perspective, is concerned with the generation of generalisable hypotheses, tenable across a range of issue areas (Rittberger 1995).[2] The empirical and conceptual work undertaken in this book, shows that relations of power have to be understood via a comprehension of what has been referred to as the problem-structure of a particular issue. The problem-structural approach is used here, for example, to show how the position of the fossil fuel lobbies is advantaged by their role in the process of energy production and consumption, so central to modern economic development. This places the lobbies on the level of a partner with governments, where the scope for policy reform is mutually determined. A similar approach is also useful in explaining the influence of IPCC Working Group 1 in policy discourses on climate change, given the level of uncertainty about the science of the issue and the scope for defining appropriate responses that this confers upon scientists.

The problem-structure of the issue simultaneously creates opportunities and puts constraints on the channels of influence open to non-governmental actors. One implication of the problem-structural approach, is that literature on global environmental politics that seeks to draw lessons from one case study in order to shed light on the politics of and potential solutions to another (Benedick 1991b; Sebenius 1991, 1992) is limited, given the distinctive nature of global warming as an issue. This draws attention to the need for detailed case study approaches to global environmental politics.

The discussion of political influence and what is implied by the term 'power' in relation to non-state actors, also provides insights that are lacking in much other work in this area. Predominant formulations of power in within IR emphasise state-based, coercive forms of power, given the historical preoccupation of the discipline

---

[2] While Wettestad (1995) refers to the 'type of problem' as an important factor affecting the effectiveness of a regime, and Young (1993) emphasises the importance of the nature of the problem being 'resolved', they do so in a way that attempts to isolate particular variables or factors of the problem-character, rather than accept the problem-structure as a set of interacting and reinforcing features.

with questions of order and conflict (see Chapter 2). Unsurprisingly, thinking on the forms of power exercised by non-governmental actors, traditionally considered irrelevant to the study of IR, is underdeveloped. Perspectives from other disciplines have enabled the development of a more comprehensive account of the question of power than is currently available in the literature on the international politics of the environment, which, given its regime orientation, is preoccupied with the power of institutions.

The utility of second-dimensional approaches to power, for example, is emphasised in all the chapters, through discussion of the importance of the anticipated reactions of actors (Crenson 1971), the existence of non-decision-making scenarios (Bachrach and Baratz 1962) and the relevance of tacit forms of power from which particular actors may benefit. They help to illustrate that inaction may also be evidence of the exercise of power; a notion missed by a focus on observable bargaining within institutions. These forms of less observable power are not explored in an explicit sense in the writing on global environmental politics, though as Chapter 2 made clear there may be some implicit subscribers to the relevance of these notions in IR more generally. Despite the methodological problems involved in making use of these approaches, it is argued that these ideas offer a useful antidote to explanations which focus on the strictly observable interactions that take place in formal policy arenas and largely between state actors, which provide only a partial account of the forms of power that operate in global climate politics.

In relation to the fossil fuel lobbies for instance, it is argued that their interests can be clearly determined even before individual policy options come to be considered, and therefore become part of government calculations, without necessarily being explicitly articulated. In other words these groups do not have to flex their political muscle, for their strength and importance in political and economic terms is already a consideration in the formulation of policy. In a similar fashion Chapter 6, on environmental NGOs, was able to show how NGOs internalise the perceived acceptability of their policy prescriptions to decision-makers, and self-censor proposals that are considered to be politically unpalatable to those they are trying to influence. This confers power upon the state, in that NGOs are inhibited from doing what they would otherwise choose to do. What is important for this analysis, is that the influence of these groups is *perceived* by decision-makers and therefore informs their policy choices, not that there is a determinable 'amount' of influence that can be attributed to each group of actors.

Addressing the issue of power in these terms also means opening up debates about the role of intention when discussing influence, and about the importance of the exclusion of agendas in international politics; debates that have provided many interesting insights in political science. This is partly a question of how the issue is constructed and understood, why some interpretations of the type of policy responses required to combat the problem of global warming are privileged over others. The book lends weight to calls by others for social constructivist approaches to environmental issues (Hannigan 1995; Litfin 1994). More imaginative conceptualisations of power are required to capture the significance of NGOs as dynamic global actors than are currently provided in the mainstream literature on global environmental politics.

In broader terms, the very fact of looking at NGOs is different from most analyses of environmental issues within the discipline of IR, which predominantly grounds itself in inter-state, institutionalist, regime analysis (Paterson 1996a; Smith 1993; Vogler 1996). Gordenker and Weiss (1995:33) note that most theories of cooperation are 'based on the state as the only noteworthy entity in international cooperation, and provide no category for considering the possibility that NGOs are significant actors in their own right'. Drawing attention to the ways in which NGOs can restrict the autonomy of states in decision-making, affect the way in which their interests are conceived and perform a number of the functions traditionally ascribed to institutions, significantly problematises many dormant and explicitly articulated ideas about international cooperation. Most problematic for regime theorists is the fact that many of the functions and impacts on state behaviour that have thus far been attributed to regimes (information exchange, deterring free-riding, stabilisation of expectations and the creation of a 'shadow of the future') can also be attributed to NGOs. In a wider sense, then, this book challenges the notion that exclusive focus on the inter-state bargaining that takes place within international institutional fora provides a satisfactory starting point for explanations of political outcome.

Opening up the issue of how international agendas come to be set as a contested domain, previously considered to be a given process where interests are assumed, also represents a step forward. This is significant for writers who are willing to attribute influence to NGOs but emphasise how participation of these groups in policy debates is always on the state's terms (Raustiala 1997; Young 1989c), because it shows how NGOs can make 'space' for their own participation by pushing and cajoling the state to act or refrain from action. The set of political relations that precede the arrival of an issue on the formal agenda of an international institution have been neglected to date, yet are shown here to reveal interesting questions about the way in which domestic and international politics interact, and how the definition of problems and the framing of appropriate responses are fought over long before the issue enters the international institutional arena.

Many claims have been made about the increasing power of NGOs in international relations, and much lip-service is paid to their importance in global environmental politics (Caldwell 1990; Hurrell and Kingsbury 1992; Levy, Keohane and Haas 1993; Porter and Brown 1991). Yet few detailed case studies of a broad spectrum of NGO activities have been undertaken. There has been little attempt, therefore, to grapple with the types of power that NGOs may have at their disposal, with what it means to say that these actors are influential. In both cases this book has served to further our understanding.[3]

---

[3] As Raustiala (1997:725) notes, although there has been more attention to the role of NGOs in recent years as 'window dressers' or guardians of the environment, or as evidence of an emerging global civil society, such characterisations 'provide limited analytic purchase on the question of the importance of NGOs within traditionally state-centric modes of international cooperation'.

7.2   Review of chapter conclusions

This review begins with the least cohesive category of non-governmental actor (and also the least explored in the IR literature): the mass media. Chapter 4 draws on media studies and other bodies of literature to show how explanations of international policy on global warming, benefit from an understanding of how the ability of actors to shape debates constitutes a particular form of power. Perhaps the most important contribution of this chapter is its attempt to extend the established notion that public opinion can be an important driving force of international action, by looking at the process by which issues become politicised and the involvement of the media in this process. The chapter builds on the idea that the way in which a problem is understood and constructed significantly affects the way in which it will be resolved; the strategies that will be employed by decision-makers to respond to the issue, and their acceptability within the public domain. It explores the processes by which particular interpretations of what global warming means as a political issue acquire salience and how others are marginalised.

It is argued that political economy, social and cultural perspectives on the operation of the media help to account for the way in which coverage is framed. The frame that currently characterises media coverage in much of the West (especially in the UK and US) is one which emphasises the scientific uncertainty surrounding the issue, the economic costs that may be implicated in policy action, and the use of technological solutions that bypass the need for structural reform. Looking at the operation of this frame helps us to understand the prevalence of a scientific discourse on climate change and the widely perceived lack of scientific consensus on the issue among the public, as well as the lack of popular recognition that the burning of fossil fuels is the major cause of the problem (Bell 1994; Rüdig 1995).

Chapter 4 posits two ways of thinking about the influence of the media in global environmental politics: a direct agenda-setting capacity, where political issues can be 'created', and a popular expectation generated that a policy response to an issue is required; and a harder to determine but possibly more influential frame, which provides a context for understanding the issue. It is this construction of the issue, repeated over time, which helps to set the boundaries of debate about what constitutes an appropriate response to the issue of climate change. By exploring the processes by which an issue reaches the political agenda, a space is opened up to explore the question of interest formation; how particular policy responses come to be required. This represents a significant advance on institutionalist accounts, which tend to disregard the process of interest formation, problem construction and domestic politics; areas that can all be informed by an analysis of the role of the media. Given the lack of attention in the IR literature to the political relevance of mass media coverage of environmental issues, this chapter offers a first attempt to redress this deficit.

Unlike the other groups of actors explored in this book, the analysis of Working Group 1 (WG1) of the Intergovernmental Panel on Climate Change (Chapter 3) is able to draw on an established body of literature, which has as its goal the assessment and explanation of the political impact of a group of knowledge-based experts (P. Haas 1990a). The ideas advanced in this literature are used extensively in Chapter 3 to account for the influence of WG1. Yet as Peter Haas (1992), is to some extent forced to acknowledge, the politics of climate change inhibit a significant degree of influence on the part of an expert scientific community, given the repercussions for policy-makers of responding to the issue, and the lack of access that has been afforded to members of the community in key administrations, such as the US. The chapter shows that while WG1 was influential in structuring the early stages of the debate on appropriate responses to climate change, and contributed to the creation of norms embodied in the text of the Framework Convention on Climate Change (FCCC), many other political factors seem to shed more light on how deals were actually brokered during the negotiations. One of the clearest indications of the influence of WG1 has been its ability to advance an agenda that emphasises the need for greater scientific research (Boehmer-Christiansen 1996), the success of which is visible in the text of the FCCC. This offers a different notion of agenda-setting influence than that which Haas accords the epistemic community.

Focus on the work of the scientific community in relation to global warming may be particularly pertinent to projecting the future of climate politics, given the emphasis by many writers upon the way in which further political action on the issue is contingent on a greater degree of scientific certainty (Skolnikoff 1990). While Chapter 3 shows that this is a simplistic assumption, the work of the IPCC can be expected to help shape the contours of future policy debates. More so than other issues, global warming is thought to be characterised by a dependence by policy-makers upon scientists to define (and legitimate) responses to the issue (ibid.).

The chapter on the influence of the fossil fuel lobbies (Chapter 5), draws on structural accounts of the way in which the position of industry lobbies in policy debates is privileged by the role they play in the creation of economic growth and the production of energy (Newell and Paterson 1998). This is used to account for the empirical evidence of the ability of the lobbies to press upon policy-makers their preferred policy choices. The forms of influence range from the drafting of parts of the FCCC to launching unprecedented lobbying efforts against the proposed European carbon tax, which succeeded in weakening the proposal in significant ways. The presence of the lobbies is argued to influence the policy formulation process through perceptions of their tacit power, their influence in policy areas not directly connected with the environment, and policy-makers' awareness of the obstacles the lobbies can place in the path of legislation that threatens their interests. Influence is also present towards the end of the policy cycle, when industry groups can scupper the implementation of proposals through strategies of non-cooperation, or by threatening to withdraw investment from countries that propose to take unilateral action.

These forms of influence are shown significantly to change the 'pay-off matrix' within which states operate, in ways that have not thus far been widely acknowledged by writers on global environmental politics. The impact of industry lobby groups has received even less attention than their environmental counterparts in this literature. The attempt in Chapter 5 to understand in a detailed way the influence of politically organised fossil fuel interests in relation to global warming, an area that has been identified as potentially prone to significant corporate influence (Sebenius 1992, 1994), goes some way towards addressing this analytical gap.

The chapter on environmental NGOs (Chapter 6), shows that whilst it is often assumed, in general terms, that environmental NGOs are responsible for putting environmental issues on to the political agenda and mobilising public opinion behind particular courses of action, the ability of groups to perform these functions varies a great deal. Particular groups from both North and South are shown to be closer to and have more influence over their national governments than others. This takes the form of an 'insider/outsider' division, where certain groups are permitted access to decision-makers and are allowed to participate on advisory committees, while others, by virtue of their lack of resources and contacts or their pursuit of a more radical agenda, find that their access is restricted. Problematically however, this does not mean that the influence of groups is, by definition, reduced by virtue of distance from the epicentre of decision-making. The case of Global Commons Institute (GCI) is used to illustrate this point. Whilst the more radical agenda being advanced by GCI means that many decision-makers are not receptive to its position, it has been able to score a few notable successes, such as forcing a reappraisal of the work of UK economist David Pearce as the basis of Working Group III's *Second Assessment Report* (IPCC 1995) and mobilising the GLOBE group and its allies within the G77 behind the notion of 'contraction and convergence'.

Environmental NGOs were also shown to be influential during the negotiation by providing advice and drafting policy proposals for 'friendly' states such as those in the AOSIS (Alliance of Small Island States) group. Yet the kinds of influence that the groups are able to exert differ significantly throughout the policy process. In general terms, the influence of environmental groups campaigning on climate change is greatest at the agenda-setting stage, where they can contribute to the translation of a problem from scientific question to policy issue, and help to raise expectations about the shape of a desired response. Their influence continues nevertheless, during the negotiating period and extends to the implementation stage, where non-compliance can be exposed and pressure brought to bear upon national policy-making.

In terms of the insights that this book can provide, by way of advancing an understanding of the political relationships at work in contemporary global climate politics, an NGO account serves to emphasise both the constraints under which states are operating in their attempts to formulate responses to global climate change, and the pressure towards action. Ascertaining the 'weight' of interests that are likely to stall action on the issue, provides an indication of what it is realistic to expect from states by way of action and the sorts of obstacle that will befall emission reduction strategies. Unless we identify the pressures that are working for and against a particular course of action, it will be difficult to assess the plausibility, in political terms, of the various strategies that are proposed for responding to the issue. Many of the obstacles that have been identified by policy analysts as standing in the way of further action on global warming, such as a lack of public concern (Hahn and Richards 1989) and the lack of a scientific consensus (Brenton 1994; Skolnikoff 1990), can be usefully informed by a perspective that is sensitive to the political influence of NGOs. It seems probable that the NGO actors examined here will continue to play a vital role in determining the nature and scope of future international policy responses to the threat of climate change.

To tie together the broader claims made in Section 7.1 (in terms of the questions this book raises and the contributions it makes in conceptual and theoretical terms) and the summary in this section of the issues raised in Chapters 3–6, the following section briefly attempts to situate the issues raised by this work in broader disciplinary terms.

7.3 The NGO project and international relations

In many ways, the approach adopted in this book is grounded in the transnationalist project of the 1970s, launched by writers such as Keohane and Nye (1977) and Nye and Keohane (1972), further developed by Willetts (1982) and showing signs of a relaunch in the late 1990s amid discussion about the need to 'bring transnational relations back in' (Risse-Kappen 1995). Much of the importance of the initial work on transnationalism was lost by writers who, when turning their attention to environmental issues, resorted to the use of a state-centred regime paradigm (Keohane 1995; Vogler 1995, 1996). In relation specifically to NGOs, Humphreys (1996:103) quotes Rosenau, who argues that 'although the early work on regimes allowed for the participation of NGOs, subsequent inquiries slipped into treating regimes as if they consisted exclusively of states'. As Risse-Kappen (1995:7) argues, the first debate on transnational relations in IR 'essentially resulted in confirming the state-centric view of world politics'. The way the transnationalist debate was first constructed made it easy for realist thinkers to dismiss the significance of non-state actors, given that states and NGOs were compared in terms of power defined by the state (Wapner 1996). In a state versus non-state actor contest, Waltz (1979) and Gilpin (1987) could claim that since NGOs could not in fact replace the state or compete on its terms, and did not possess the same forms of power, they were irrelevant.

This book may be able to lend some conceptual and empirical weight to the renewed interest in transnational relations in the late 1990s by showing how non-state actors can structure the environment in which states operate, and can themselves use the interconnections between domestic and international politics (of which they are a part) to advance their agendas.[4] Hence the emphasis on non-state actors, the complex interdependencies that exist between non-state and government actors and the global (as opposed to domestic or international) scope of the project, would be appealing to the transnationalist school. By discussing forms of non-state

---

[4] Transnational relations are defined (narrowly) by Risse-Kappen (1995:3) as 'regular interactions across national boundaries when at least one actor is a non-state actor or does not operate on behalf of a national government or an intergovernmental organisation'.

power the book may also help to develop a response to Turner's (still valid) criticism of the intellectual response to transnationalism: while 'Nye and Keohane have undoubtedly helped steer foreign policy specialists away from a too rigid conception of the central role of government bodies ... what they have been less successful in illuminating is the extent of the power such transnational actors have been able to exert' (Turner 1978:16–17).

This book also draws on established critiques of the lack of attention in writing on IR to the significance of domestic politics in explaining international events (Putnam 1988). In order to do this, classic literatures on bureaucratic politics have been used (for example Allison 1971) to emphasise the fragmented and non-unitary nature of the state. In all the chapters of this book, emphasis is laid upon the ways in which (particularly at the agenda-setting and implementation stages) domestic events structure the international political milieu. The fragility of the division between domestic and international politics, however, is particularly exposed by an NGO reading of the politics of climate change, where connections between the two levels are shown to be multiple, complex and interactive. In particular, Chapters 5 and 6, which emphasise the global reach of the fossil fuel lobbies and environmental NGOs, help to go beyond the two-level game (Putnam 1988) – which shows how the domestic and international interact – by drawing attention to the global level (beyond the inter-state) and how this level interacts with all the others.[5] In this respect the book illustrates how the global is as likely to structure the domestic (and international) as the other way around.

Building on this, the empirical evidence presented here highlights the need to get 'inside' the state, to show how NGOs can enjoy influential relations with particular parts of government, and to the extent that those departments influence the overall direction of government policy, NGOs can extend their influence in important ways. For example, the close relations between the fossil fuel industries and Departments of Energy, Trade and Industry are important, because of the influence these departments have been able to exercise over governments' climate policy.

A further point that emerges from this discussion of factors within and beyond the state, is the need to explore the ways in which the politics of decision-making in non-environmental areas affect policy-making on the environment. Evidence of what was referred to as 'cross-issue influence' shows how the influence of non-state actors across a range of policy areas affects power relations within the traditionally and narrowly defined area of environmental policy-making. The preoccupation of regime analysis with issue-areas excludes other forms of explanation based on inter-actions outside the regime. This idea is particularly developed in the chapter on the fossil fuel lobbies. The influence of the latter in transport and energy policy restricts the effectiveness of policy options designed to combat climate change, and reduces governments' room for policy manoeuvre.

---

[5] This extension is broadly compatible with Risse-Kappen's (1995:300) suggestion of a three-level game, where the third level is the 'transnational'.

It should be acknowledged, however, that the approach taken in this book defines NGOs in terms of their ability to change the behaviour of states (directly and indirectly), whereas to assess the wider significance of the various NGOs as global actors, their broader social, economic and cultural roles would have to be considered.[6] In this sense the book does not lend much to the conceptualisation of the NGO phenomenon as a political development in its own right. The way in which NGOs tend, in this analysis, to be defined by their relations with the state is on one level ironic, given that the book seeks to move away from traditional state-centred approaches. In this sense little has been done to rectify Wapner's (1996:10) charge that most studies of NGOs emphasise their importance only to the extent that they 'change states' policies or create conditions in the international system that enhance or diminish inter-state cooperation'. There is a danger inherent in an enquiry that attempts to challenge state-centric explanations of global environmental politics by focusing on how NGO activities change state interests and strategies, that state-centric accounts are reproduced. Yet the imperative of demonstrating to IR scholars of the environment the significance of NGOs, means in the first instance developing an approach in which the state retains a key role in order to demonstrate how, to achieve their own research goals, regime writers have to take account of NGOs as actors, because they can significantly change the ways in which we seek to explain state behaviour. So whilst such a project may be cast 'in the shadow of the state' (Wapner 1996:162), as Evans argues, 'it is not interesting to exclude traditional state behaviour and then study the residual only' (quoted in Nye and Keohane 1972:24). Risse-Kappen (1995:13) notes, moreover, that 'One can subscribe to the proposition that national governments are extremely significant in international relations and still claim that transnational actors crucially affect state interests, policies and inter-state relations . . . one does not have to do away with the "state" to establish the influence of transnational relations in world politics.'

This emphasis means, however, that the project remains part of a strong tradition in IR of seeking to explain the outcomes of political interactions in formal institutional fora, although the argument here is that the ways in which this can be done would benefit from a less state-centred, NGO perspective. Hence whilst this book might be understood as a reaction to regime perspectives, by engaging with these accounts and seeking to explain the same outcomes differently, it shares the principal goal of the regime project: to account for cooperation at the international level. The framework adopted here challenges the ability of these approaches fully to explain policy outcomes, but it does so in a way that may usefully complement their stated aim of explaining regime formation, maintenance and change.

This is not to suggest, however, that this account can be sidelined as a footnote to regime perspectives or an appendix to work on global warming. The approach adopted here shows that existing forms of explanation are in need of reformulation

---

[6] Princen and Finger (1994) and Wapner's (1995) work on NGOs as transnational actors and Kuehls' (1996) work on NGOs as 'transversal' actors are examples of writers who have sought to engage with the significance of these actors in broader political and social terms.

if they are effectively to achieve their stated goals. If regime writers are to sharpen their ability to account for the global politics of climate change they must address the significance of NGOs as political actors. It is not just a question of NGOs performing, enhancing or undermining functions traditionally performed by international regimes, as was argued above, but of NGOs contributing to states' assessment of their interests and the collective definition of problems, that makes an NGO framework centrally relevant to the regime project. Regime approaches to global environmental change, if they are to avoid the fate that Smith (1993) foresees for them (marginalisation within the discipline of IR), must embrace wider accounts of global politics than are permitted by strictly institutionalist, inter-state analysis. In this respect it is hoped that this book will help to challenge the 'boundaries and assumptions' (ibid.:40) of the study of environmental change in IR.

The book has categorically not sought to establish where the balance of power lies between the state and non-governmental actors. The question that has been addressed here is not the extent to which the state is more powerful than various non-governmental actors, or the degree to which these groups are more powerful than the state. The power assets that states and non-state actors have at their disposal are in any case largely incomparable in a meaningful way. This may disappoint the realist reader looking to be proved wrong that states remain the primary actors in international society. It would be a most unhelpful exercise to compare the power of the state with that of NGOs in terms defined by forms of state power. This would tell us little about the significance of NGOs. Their influence in many ways depends on their ability to mobilise assets and resources that other actors do not have; this is why they are able to influence other actors. Nor should the argument be used to sustain claims about the 'retreat of the state' (Strange 1996) or of a power shift in global politics. The analysis provided here is of a micro set of relations peculiar to a particular case study and therefore could not be used to support such broader claims. Whilst there may be evidence of a realignment of responsibilities between states and non-state actors, the emphasis is on the webs of influence that enmesh state and non-state worlds and not a zero-sum transfer of power between them.

It is possible to make the case that NGOs are influential actors in their own right and still concur with the argument that 'Rather than undermining state sovereignty, active NGO participation enhances the abilities of states to regulate globally' (Raustiala 1997:719). Oppositional depictions where an increase in the power of an NGO must necessarily come at the expense of state power are misleading, as the significance of non-state actors should not be reduced to the extent to which they strengthen state capacity. Similarly, the problem with analysis that seeks to challenge claims about NGO influence by referring to the enduring power of the state to set the context within which they operate, is that it inevitably serves to confirm realist suspicions that NGOs only matter when they matter to states, when they exercise state-like power. In this formulation power is reduced to a statist concept. The claim that because 'states remain in control of the process' and set the parameters of participation, they 'gain advantage from NGOs participation' (quoted in Ringius 1997:68) overlooks the fact that NGO influence is extended when the states take advantage of the services that NGOs provide. Grubb et al. (1999) argue that the negotiations culminating in the Kyoto Protocol show that Kyoto was very much an agreement struck by states, negotiating what they felt to be possible and appropriate in the face of strong opposition from powerful industries and including flexible mechanisms that were opposed by most environmental NGOs. The analysis is used to reaffirm the pre-eminence of states in the process. This is achieved, as in Arts' (1998) work, by assessing the impact of NGOs upon the formal negotiations and implementation of the protocol. Their role in bringing it about in the first place is not considered. The process of nurturing expectations and structuring the debate at the domestic level is downplayed. Such constructions of the legitimate and relevant domain of climate politics assume a very narrow definition of influence that privileges the study of 'vacuum politics', where negotiations become a laboratory in which those on the inside are immune from outside concerns and pressures.

The non-state actors that are the subject of this book are not powerless because they do not draft the final details of legal documents, or because they are not in a closed meeting that decides upon the exact wording of a protocol text. They are powerful because they have helped to shape the contours of the debate by lobbying for or against various policy options, by projecting new understandings, norms and expectations, or by helping to ensure that some options were never even considered because of their structural power, which may not be visible in the confines of the formal policy arena but is ever-present in the minds of policy-makers. These sources of influence derive from their ability to negotiate the multilayered webs of activity from the national to the regional and global. This power stems not from a direct role in the formal bargaining that takes place in the halls of the UN during episodic gatherings on environmental matters as such, but from claims to social authority supported by society and from economic might wielded in the day-to-day functioning of the global economy.

The range of approaches adopted throughout the book to explain NGO influence cannot easily be pigeonholed into one of the pre-existing theoretical paradigms in IR. They resonate with the goals and emphases of a number of strands of thinking in the discipline, and could be used to support a variety of, potentially conflicting, contentions. That the types of explanation advanced here do not obviously sit comfortably within mainstream explanations of international politics, is perhaps evidence of the failure of academic enquiry to date to incorporate the dynamism and relevance of non-state actors into constructions of global politics, and highlights the need for approaches that claim to be able to explain global environmental politics, to refocus their conceptual lenses.

Hence although this book has not explicitly sought to compare the approach of each strand of IR thinking to the study of global climate politics, it is clear that the account of global climate politics offered here is not easily captured by predominant regime perspectives. It is apparent not only from the ground covered in this book, but also in the conclusions drawn by Rowlands (1995) and Paterson (1996b), that traditional IR theory can only inadequately capture the political dynamics of the climate change issue. There is clearly space for a differently grounded NGO account.

7.4   Towards an alternative perspective

An alternative framework would need to be able to explain international outcomes as the product of a policy cycle from agenda-setting through to implementation in a way which is sensitive to the interplay between different 'levels' of analysis from domestic to global. It would have to capture power relations, not as they manifest themselves in one particular environmental issue area, but as part of a wider ensemble of structures and webs which enable and constrain political opportunities. Given the political and economic context in which climate change is addressed, this means locating the approach within the context of an increasingly global political economy.

It is impossible to make a strong case for an alternative approach without applying it to a case study. Clearly such an endeavour is beyond the scope of this book and would require another book-length project. What this final section will do, however, is suggest the possible merits of a *political economy of transnationalism*. This would be a transnationalism that is sensitive to the state's location in circuits of power and networks of interdependency that bind it to a multiplicity of actors and institutions that significantly shape its field of operation. Whilst it was suggested above that the approach developed in this book lends broad support to the claims of transnationalists such as Risse-Kappen (1995), the importance of political economy, which emerges strongly both from this book and the work of Matthew Paterson (1996b), suggests the need to reconfigure transnationalism within a political economy framework.

Risse-Kappen (1995:17) clings to a separation between state and social forces, which is not appropriate. He is keen to retain an independence for the state from 'societal interest groups' and refers to circumstances in which the state is able to 'win' against transnational actors, implying the state's abstraction from these transnational relations. He distinguishes between transgovernmental relations (ibid.:9), which are viewed as a transnational equivalent of bureaucratic politics (the state sphere), and transnational relations (the NGO sphere), rather than explore how the two overlap and interrelate. A reconfigured transnationalist framework would offer a multi-level analysis of the conditions in which transnational actors make an impact which gets inside the state and looks at transgovernmental coalitions and transnational alliances as part of the same phenomenon.

An account that seeks to identify cross-sectoral, trans-state coalitions that include particular parts of government and international organisations in alliance with different interest groups, would help to focus attention on the capillaries of power that run through the global system. It would allow us to capture loose clusters of interests including, for instance, the Australian government and elements of a Republican-dominated US Congress, operating alongside US fossil fuel industries, climate sceptics in the scientific community and conservative media; all working to prevent action on climate change. At the same time 'sunrise' (renewable energy) industries, environment departments, 'activist' scientists and environmental groups would constitute a loose advocacy coalition, sometimes operating in an overtly coordinated fashion and at other times on a more tacit basis. Such clusters coalesce around policy positions at the national and the international level, grounded in a shared recognition of mutual interests. This helps to problematise the distinction between non-state and state actors; to challenge further the idea of the state as a homogeneous unit and NGOs as strictly non-state actors. It is clear from this book that there can be no hard and fast distinctions, when representatives from environmental and industry groups sit on government delegations and draft negotiating texts, and when scientists work for governments in Environment Ministries. The world of climate politics is clearly not as conveniently delineated between domestic/international, and state/ non-state, as many readings imply.

Such an approach would also help to get beyond the tendency to view dimensions of NGO power in isolation from one another, and instead look at the way in which they interact and mutually reinforce one another. In her analysis of the positions adopted by India and Brazil in the climate change negotiations, Jakobsen (1997:13) argues, that 'A web of domestic and foreign experts, NGOs and international opinion seems . . . to have framed the contours within which Brazilian and Indian bureaucrats have chosen one policy option over another.'

Her work supports the contention here that the resolution of the climate change issue will be determined by the interactions between various parts of the state, social movements and the power brokers in the global economy. The state will not be the mediator of all (or even most) of these transactions and conflicts. We should be as alert to the shifting coalitions of NGO, expert community and government department interests as we have traditionally been to the manoeuvrings of negotiating blocks such as OPEC and AOSIS. The current of political change in the global warming debate is only partially manifested in the visible manoeuvrings of governments in international fora. The history of the climate change issue has always been one of competing authority claims and a plurality of voices. The challenge is to acknowledge the diversity of actors and interests without assuming open decision-making processes or the absence of mobilised bias.

7.5 From transnationalism to IPE

It is becoming increasingly clear that the world of environmental politics is not insulated from the influences and constraints associated with global economic change that affect other areas of policy. Therefore, a transnationalist framework for comprehending networks of NGO influence would also benefit from insights from international political economy (IPE). An IPE perspective would allow us to look at the economic ties and interdependencies that underpin transnational policy alliances. Paterson (1996b) articulates an IPE of climate change using a perspective rooted in historical materialism to understand the North–South elements that are central to the climate change debate, and to look at the importance of globalisation and the shifting patterns of production and how these work to constrain or enable state autonomy in areas of environmental decision-making. The approach suggested here – whilst similar in terms of the use of the state theory with particular reference to the state's role in the energy business (ibid.:160) – would use work within IPE to ground an analysis of transnational relations. In particular, it would help to understand the privileging of particular actors in the climate change debate. The chapter on the political influence of the fossil fuel lobbies (Chapter 5) highlighted the importance of state theory; of comprehending the function of the state in a capitalist world economy as a basis for understanding the biases of the state towards some lobbies (Newell and Paterson 1998). This does not mean that state behaviour can be predicted by assessing the net demands of capital-in-general. The interests of capital are diverse and particular factions compete to present their interests as those of capital-in-general. Perceptions about which sections of capital serve the national interest will vary over time. Hence although the importance of the fossil fuel industries to the production of energy in contemporary industrialist societies means that this sector shares key interests with the state, the sensitivity of the insurance industries to the effects of climate change (Brown 1996c, Leggett 1995b) means that in the future the power of an important sector of finance supporting action on climate change may be brought to bear upon the state (Newell and Paterson 1998). A political economy perspective would also be able to show how many other actors important to the climate change debate are inserted into the patterns and structures of the global economy. It would understand the role of the media in framing the public debate on climate change in such a way that the role of industries in producing climate change is obscured as a function of the ties the media has to the fossil fuel industries through advertising revenue. It would also help to illuminate aspects of the relationship between scientists and policy-makers where reliance by the former upon funding from the latter shapes the nature of interaction between the two communities and the nature of the advice which scientists offer.

A political economy approach to the study of climate politics can contribute a great deal to an understanding of the context of cooperation. It may guide expectations about the sorts of policy response that are likely to be forthcoming. Shifting relationships within the global economy will have immense repercussions for the future of global climate politics, and are likely to change the conduct of international relations. As Susan Strange (1994) argues, trilateral diplomacy is a growing feature of world politics, where firms bargain with other firms and states in a triangular fashion. In the climate context, joint implementation and tradable permit schemes will require significant inter- and intra-firm cooperation as well as state–firm cooperation. States may help to set down the rules of the game, but the success and nature of the policy instrument will largely be a matter of inter-business relations. That such arrangements are considered necessary is simultaneously a recognition of the centrality of investment decisions to the success of climate change efforts and of governments' desire to avoid policy measures which negatively impact upon business. The privileging of market solutions over state-based interventions is endorsed and legitimised through the ideology of neoliberalism (Gill 1994). Those that deliver such solutions become key players, creating a space for the primacy of strategies promoted by industry.

What this implies is the need for a political economy of transnationalism that goes beyond emphasising the diversity of actors in international society and the interconnections between them, and takes a broader view of the processes and structures, of which non-state actors are a part, that shape the contours of international cooperation. To improve upon existing accounts, such an approach would have to address the importance of knowledge, its construction and diffusion. This book has shown, for example, how science is used to validate the claims of all actors in the debate, from environmental and industry groups to government officials and the media. The role played by the media and scientific community in framing and acting as gatekeepers for what is considered relevant public and policy information has been shown to be key to understanding the boundaries and assumptions at work in contemporary climate politics. Although a 'constructivist political economy' is still underdeveloped (Paterson 1996b:180), it is reasonable to expect that a political economy perspective could help us to understand the role of knowledge in environmental politics. For example, the process by which particular understandings of the issue acquire salience and are legitimised, and how others are marginalised, would be usefully informed by a political economy approach. And whilst some important aspects of the role of knowledge in climate politics would not be captured by IPE perspectives, reference to political-economic contexts would provide an appreciation of why debates are constructed in a particular way, why – despite significant contestation – particular understandings are upheld and institutionalised. Social understandings and knowledge constructions reflect the economic as well as the social and cultural values of those in a position to frame debate and shape policy discourse. Frames are sponsored by particular interests in the debate and the extent to which they resonate with those who make the final decisions is inevitably a function of their economic power.

One other way of linking institutions, ideas and material capabilities is to draw on the work of Robert Cox. Levy and Egan (1998), for example, root Cox's work more broadly within a neo-Gramscian framework in order to explore the dimensions of corporate power in the climate change debate. They look at the mutually supportive interaction of discursive and structural power, where media ownership structures complement the structural power of business in the economy, which in turn rests (in part) on discourses that associate general prosperity with corporate profitability. A Coxian approach may help to show how particular configurations of social forces and state forms, expressed through specific institutional formations, help to produce and maintain consensus around a restricted set of understandings about appropriate responses to climate change. It could be argued that media coverage of the climate issue reflects and reinforces the prevailing production structure, which is heavily dependent on the use of fossil fuels. This production structure also constrains state decision-making and the scope of international institutional action on the issue. The state has also, of course, actively promoted such a structure of production by providing large subsidies and tax breaks, as well as promoting understandings of the climate change problem which do not implicate this production structure in causing it.

A Coxian approach may be useful in another sense. The notion of a 'managerial class' (Cox 1987)[7] resonates with the interpretations of global environmental politics offered by Chatterjee and Finger (1994), Hildyard (1993), and Sachs (1993), and can be used to develop a political economy of transnationalism by exploring the strategic and tacit clusters of interest described above. In explaining UNCED, therefore, attention would focus on how the boundaries and assumptions of debate have been drawn and sustained in a way that accommodates challenges and threats to the legitimacy of power brokers, such that it extends and entrenches the power of those policy networks that are faced with addressing the problem. Whilst for the global ecology writers cited above, élitism and insider/outsider formulations are the starting point for understanding these coalitions of interest, for Cox the emphasis is more upon class (though rivalries exist within this class and its members do not necessarily identify themselves as being part of a class), just as clusters of interests in the climate debate do not identify themselves as part of a broader coalition.

It is helpful, then, to view cooperation and the institutions that promote it as being the product of a particular ensemble of power relations. Transnational coalitions of interest help to institutionalise (but are also served by) a particular mobilisation of bias. In seeking to explain, therefore, why particular transnational actors make an impact, a political economy of transnationalism would require us to do more than look at domestic structures and the mediating impact of international institutions (as Risse-Kappen does). Instead you would have to ask why some coalitions are privileged in the debate to the active exclusion of others. This requires an appreciation of the role of these alliances in the global political economy; the way in which internationalisation shapes the scope of policy action and brings about change in the patterns of domestic and international influence of particular interests (Keohane and Milner 1996). For example Cox's (1994:357) emphasis on the 'internationalisation of the state' shows how precedence is given to certain state agencies, such as Ministries of Finance and Trade, because of their key role in responding to international economic imperatives. This gives rise to a 'new axis of influence' (ibid.), where international policy networks are linked with the central agencies of government and big business. The issue, then, is not state versus NGO or international versus domestic, but (often) tacit clusters of actors and interests that benefit from the prevailing structure of economic and political organisation, which they use to legitimate their role in shaping responses to the problem. This is useful in understanding why some NGOs are more influential than others, why some parts of government exercise more power than others, and how when each seeks to extend their own influence, they find it convenient to forge associations with one another.

---

[7] Cox's definition of a managerial class 'encompasses public officials in the national and international agencies involved with economic management and a whole range of experts and specialists who are in some way connected with the maintenance of the world economy in which multinationals thrive' (Cox 1987:360–1).

A Coxian approach can only lend generic insights to a project that is oriented towards explaining particular outcomes and engages so explicitly with the question of influence at a particular historical moment. In order to take a Coxian analysis further, would mean placing the climate change debate in a much broader historical analysis, one rooted in the changing patterns of production and structure associated with capitalist development, rather than the 'actor-interactions paradigm' within which this project is located (ibid.:395). The broad approach, nevertheless, suggests one starting point for more effectively integrating political economy into the study of the transnational organisation of the state and social forces.

Hence whilst applauding the attempt by Risse-Kappen (1995) and others to inject new life into the transnational relations/complex interdependence school, it is suggested here that such an approach would provide a more convincing account of the global politics of the environment – if it were to integrate more fully transgovernmental and transnational relations as well as shed some of its pluralist underpinnings. Explanations that fail to situate networks of transnational relations within the global political economy of which they are a part, will only ever provide a partial understanding of the interests and politics at the heart of environmental cooperation and degradation. What a Coxian approach seems to offer is the necessary integration of an IPE approach with an understanding of the mutually reinforcing relationship between the state (form) and social forces.[8]

7.6   Conclusion

This book has suggested a number of ways in which non-state actors are important to explanations of the international community's response to the threat of global climate change. The empirical story sketched above only provides a snapshot of the current landscape of NGO activity in global climate politics, and future changes will need to be continually documented. This is especially important in an area of politics so subject to flux and conflict as global warming; power relations will transform and realign as circumstances change.

There is some indication that the regime thinkers who currently dominate the discourse on international environmental politics, recognise the necessity of future research into non-state actors. As Young and Von Moltke (1994:361–2) argue, 'it is critical to deepen our concerns for the pervasive role of non-state actors as players in the processes of regime creation, implementation and operation of regimes. . . . No issue-area constitutes a better laboratory in which to study these developments than international environmental affairs.' It is hoped, then, that this book has been able to demonstrate, not just that the existing approaches to explaining global climate politics provide an incomplete representation of the political relations that surround

---

[8] For Cox (1987:105) the principal characteristics of state forms are defined by the characteristics of their historical blocs, that is, 'the configurations of social forces upon which state power ultimately rests. A particular configuration of social forces defines in practice the limits or parameters of state purposes'.

this complex issue area, but that most of the approaches outlined may constitute the foundations of an alternative framework for thinking about the importance of non-state actors in global environmental politics. At minimum they suggest a bridge from the predominantly state-based literature to a wider account of transnational world politics, in which the importance of non-state actors is better represented.

APPENDIX A

List of abbreviations

ABARE	Australian Bureau of Agriculture and Resource Economics
AGBM	Ad Hoc Group on the Berlin Mandate
AGGG	Advisory Group on Greenhouse Gases
AIJ	Activities implemented jointly
AIM	Accuracy in the Media
ALTENER	Programme for the promotion of renewable energy sources (EU)
AOSIS	Alliance of Small Island States
APPEA	Australian Petroleum Production and Exploration Association
BBC	British Broadcasting Corporation
BCAS	Bangladesh Centre for Advanced Studies
BCSD	Business Council for Sustainable Development
BDI	Bundesverband der Deutschen Industrie
BRF	British Roads Federation
BTU	British thermal unit
CAFE	Corporate average fuel efficiency standards
CAN	Climate Action Network
CANLA	Climate Action Network Latin America
CANSA	Climate Action Network South Asia
CANSEA	Climate Action Network South East Asia
CASA	Citizen's Alliance for Saving the Atmosphere and Earth
CBI	Confederation of British Industry
CDM	Clean Development Mechanism
CEA	Council of Economic Advisors (US)
CFC	Chlorofluorocarbon
CNA	Climate Network Africa
CNE	Climate Network Europe
CO_2	Carbon dioxide
COP	Conference of the Parties
CSE	Centre for Science and Environment (India)
DoE	Department of Energy (US)
DoT	Department of Transport
DTI	Department of Trade and Industry
EBCSEF	European Business Council for a Sustainable Energy Future
EC	European Community
ECJ	European Court of Justice

ECOSOC	Economic and Social Council (UN)
ECSC	European Coal and Steel Community
EDF	Environmental Defense Fund
ENDA	Environment and Development Third World
EPA	Environmental Protection Agency
ERT	European Roundtable of Industrialists
EU	European Union
EUROACE	European Association for the Conservation of Energy
EUROPIA	European Petroleum Industry Association
FCCC	Framework Convention on Climate Change
FIELD	Foundation for International Environmental Law and Development
FoE	Friends of the Earth
FoEI	Friends of the Earth International
GCC	Global Climate Coalition
GCI	Global Commons Institute
GCMs	General circulation models
GEF	Global Environmental Facility
GLOBE	Global Legislators for a Balanced Environment
GNP	Gross national product
G7	The seven most industrialised countries in the world
G77	The group of seventy-seven less developed countries
HFC	Hydrofluorocarbons
ICC	International Chamber of Commerce
ICCP	International Climate Change Partnership
ICSU	International Council of Scientific Unions
IEA	International Energy Agency
IFIEC	International Federation of Industrial Energy Consumers
IGBP	International Geo-Biosphere Programme
INC(CC)	Intergovernmental Negotiating Committee on Climate Change
INGO	International non-governmental organisation
IPCC	Intergovernmental Panel on Climate Change
IR	International relations
IUCN	International Union for the Conservation of Nature
JI	Joint implementation
JOULE	Programme on Joint Opportunities for Unconventional or Long-term Energies
JUSCANZ	Negotiating bloc made up of Japan, New Zealand, Australia and the United States
LDCs	Less developed countries
MOEF	Ministry of Environment and Forests (India)
MITI	Ministry of International Trade and Industry (Japan)
NASA	National Aeronautics and Space Administration
NCA	National Coal Association

NGO	Non-governmental organisation
NIEO	New International Economic Order
NRDC	Natural Resources Defense Council
OECD	Organisation for Economic Cooperation and Development
OMB	Office of Management and Budget
OPEC	Organisation of Petroleum Exporting Countries
PAC	Political Action Committee
PAMs	Policies and measures
PFCs	Perfluorocarbons
QELROs	Quantifiable emission limitation and reduction objectives
SAR	Second Assessment Report of the IPCC
SAVE	Specific Actions for Vigorous Energy Efficiency
SBI	Subsidiary Body on Implementation
SBSTA	Subsidiary Body for Scientific and Technical Advice
SEI	Stockholm Environment Institute
SF_6	Sulphur hexafluoride
SMMT	Society of Motor Manufacturers and Traders
SWCC	Second World Climate Conference
TERI	Tata Energy Research Institute
THERMIE	Energy Technology Support Programme
UK	United Kingdom
UN	United Nations
UNCED	United Nations Conference on Environment and Development
UNCTAD	United Nations Conference on Trade and Development
UNECE	United Nations Economic Commission on Europe
UNEP	United Nations Environment Programme
UNESCO	United Nations Educational, Scientific and Cultural Organisation
UNFCCC	United Nations Framework Convention on Climate Change
UNGA	United Nations General Assembly
UNICE	Union of Industrial Employers' Confederations in Europe
UPI	Italian Oil Board
US	United States of America
WBCSD	World Business Council for Sustainable Development
WCI	World Coal Institute
WCRP	World Climate Research Programme
WEC	World Energy Council
WG	Working Group (of the IPCC)
WICE	World Industry Council for the Environment
WMO	World Meterological Organisation
WRI	World Resources Institute
WWF	World Wide Fund for Nature/World Wildlife Fund (US)
WWFI	World Wide Fund for Nature International

APPENDIX B

Chronology of the international response to the issue of global climate change

1979

February   Geneva (Switzerland) First World Climate Conference
September–October   Villach (Austria) workshop
November   Bellagio (Italy) meeting to follow up Villach workshop

1988

June   The 'Toronto Conference on the Changing
                                    Atmosphere: Implications for Global Security'
                                    proposes a 20 percent reduction in CO_2 emissions by
                                    the year 2005, taking 1988 as a base year. World
                                    Meteorological Organisation and United Nations
                                    Environment Programme establish the
                                    Intergovernmental Panel on Climate Change (IPCC)
November   First meeting of the IPCC, Geneva

1989

February   Legal aspects of climate change discussed by experts,
                                    Ottawa (Canada)
March   Summit meeting on the protection of the atmosphere,
                                    the Hague (Netherlands)
November   Ministerial Conference of Atmospheric Pollution and
                                    Climate Change, Noordwijk (Netherlands)

1990

April   White House Conference on Scientific and Economic
                                    Research related to Global Change, Washington
May   UNECE (United Nations Economic Commission on
                                    Europe) meeting to respond to the WCED report,
                                    Bergen (Norway)
August   IPCC meeting to approve First Assessment Report,
                                    Sundsvall (Sweden)
November   Second World Climate Conference, Geneva

1991

	INC (Intergovernmental Negotiating Committee (on Climate Change)) meetings:
4–14 February	INC1
19–28 June	INC2, Geneva
9–20 September	INC3, Nairobi (Kenya)
9–20 December	INC4, Geneva

1992

18–28 February	INC5 New York
30 April–9 May	INC6, New York
7–10 December	INC7, Geneva
June	UNCED Summit, Rio de Janeiro (Brazil). Framework Convention on Climate Change opened for signature

1993

15–20 March	INC8, New York
16–27 August	INC9, Geneva

1994

7–18 February	INC10, Geneva
March	The convention enters into force
22 August– 2 September	INC11, Geneva

1995

6–17 February	INC12, New York
28 March–7 April	COP1, Berlin (Germany)
21–5 August	AGBM (1st session), Geneva
28 August– 1 September	SBSTA (1st session), Geneva
31 August– 1 September	SBI (1st session), Geneva
30 October– 3 November	AGBM (2nd session), Geneva
30–1 October	AG13 (1st session), Geneva

1996

27 February– 4 March	SBI and SBSTA (2nd session), Geneva
5–8 March	AGBM (3rd session), Geneva
June	IPCC Second Assessment Report released.
8–19 July	AGBM (4th session), Geneva
8–19 July	COP2
9–16 July	SBI and SBSTA (3rd session), Geneva
10 July	AG13 (2nd session), Geneva

10–11 December	SBI (4th session), Geneva
16–18 December	SBSTA (4th session), Geneva
9–13 December	AGBM (5th session), Geneva
16–18 December	AG13 (3rd session), Geneva

1997

24–8 February	SBI and SBSTA (5th session), Bonn
25–8 February	AG13 (4th session), Bonn
3–7 March	AGBM (6th session), Bonn
3–12 June	AG13 (6th session), Bonn
28 July–7 August	AG13 (5th session), Bonn
28 July–7 August	SBI and SBSTA (6th session), Bonn
31 July–7 August	AGBM (7th session), Bonn
20–9 October	SBI and SBSTA (7th session), Bonn
22–31 October	AGBM (8th session), Bonn
1–12 December	COP3, Kyoto (Japan). Kyoto Protocol opened for signature.

1998

2–12 June	SBI and SBSTA (8th session), Bonn
2–13 November	SBSTA (9th session), Buenos Aires (Argentina)
2–12 November	COP4, Buenos Aires
3–10 November	SBI (9th session), Buenos Aires

1999

31 May–11 June	SBI and SBSTA (10th session), Bonn
25 October–	
5 November	COP5, Bonn

References

Achanta, A. (ed.) (1993) *The Climate Change Agenda: An Indian Perspective*. New Delhi: TERI.

Achbar, M. (1994) *Manufacturing Consent: Noam Chomsky and the Media*. New York: Black Rose Books.

Acid News (1994a) 'Global warming factories', no. 2, April.

Acid News (1994b) 'Power plants pilloried', no. 2, April.

Acid News (1994c) 'Leading up to Berlin', no. 2, April.

Acid News (1994d) 'Negotiations stalled', no. 4, October.

Acid News (1994e) 'Efficiency standards halted by controversy', no. 5, December.

Acid News (1995a) 'Not very interested', editorial, no. 2, April.

Acid News (1995b) 'Infrastructure above all', no. 4, October.

Acid News (1996) 'Voluntary commitment', no. 3, June.

Adams, W. (1982) *Television Coverage of International Affairs*. New Jersey: Ablex.

Adler, E. and Haas, P. M. (1992) 'Conclusion: epistemic communities; world order and the creation of a reflective research program'. *International Organisation*, 46(1), pp. 367–90.

Advisory Committee on Business and the Environment (1998) Climate change: a strategic issue for business', 31 March. London: DETR.

Agarwal, A. and Narain, S. (1990) *Global Warming in an Unequal World: A Case of Eco-colonialism*. New Delhi: Centre for Science and Environment.

Agence Europe (1995) EU Tax on CO_2 emissions: 'MEPs unhappy with the absence of results three years after the Earth Summit – The Council President admits the need for fiscal measures'. Brussels, plenary session of the European Parliament, 1 April.

Aguilar, S. (1993) 'Corporatist and statist designs in environmental policy: the contrasting roles of Spain and Germany in the EC scenario'. *Environmental Politics*, 2(2), Summer, pp. 223–47.

Akumu, G. (1996) Climate Network Africa. Interview, COP2, Geneva, 17 July.

Allen, S. (1997) 'Are skeptics winning the debate on warming?'. *Boston Globe*, 28 April.

Allison, G. (1971) *Essence of Decision: Explaining the Cuban Missile Crisis*. Boston: Little Brown.

Anderson, A. (1991) 'Source Strategies and the communication of environmental affairs'. *Media, Culture and Society*, 13(4), pp. 459–76.

Anderson, A. (1993) 'Source–media relations: the production of the environmental agenda'. In A. Hansen (ed.) *The Mass Media and Environmental Issues*. Leicester: Leicester University Press, pp. 51–68.

Anderson, A. (1997) *Media, Culture and the Environment*. London: UCL Press.

Anderson, A. and Gaber, I. (1993a) 'The yellowing of the greens'. *British Journalism Review*, 4(2), pp. 49–53.

Anderson, A. and Gaber, I. (1993b) 'Antisappointment: The Rio summit and the coverage of environmental issues on British television news'. Conference paper (conference title unavailable).

Anderson, D. (1995) 'Rapporteur's report of workshop presentations and discussions'. In M. Grubb and D. Anderson (eds) *The Emerging International Regime for Climate Change*. London: R11A, pp. 7–42.

Anderson, R. (1996) Research Manager, American Petroleum Institute, Questionnaire on the Politics of Global Warming.

Andresen, S. (1989) 'Increased public attention: Communication and polarisation'. In S. Andresen and W. Ostreng (eds) *International Resource Management*, pp. 25–46.

Andresen, S. (1991) 'US Greenhouse Policy: Reactionary or Realistic?'. *International Challenges*, 11(1), pp. 17–24.

Andresen, S. (1992) 'The Climate Negotiations: Lessons and Learning'. *International Challenges*, 12(2), pp. 34–42.

Andresen, S. and Ostreng, W. (1989) (eds) *International Resource Management: The Role of Science and Politics*. New York and London: Belhaven Press.

Andresen, S. and Wettestad, J. (1990) 'Climate Failure at the Bergen Conference?'. *International Challenges*, 10(2), pp. 17–23.

Andresen, S. and Wettestad, J. (1992) 'International resource cooperation and the greenhouse problem'. *Global Environmental Change*, 2(4) (December), pp. 277–91.

Andresen, S., Skodvin, T., Underdal, A. and Wettestad, J. (1994) 'Scientific management of the environment? Science, politics and institutional design'. *International Challenges*, 14(3), pp. 9–13.

APPEA (Australian Petroleum Production and Exploration Association Limited) (1996) 'Cooperative agreement between APPEA Ltd and the Commonwealth of Australia in relation to action to abate emissions of greenhouse gases'. Press release, COP2, Geneva, July.

Arp, H. (1993) 'Technical regulation and politics: The interplay between economic interests and environmental policy goals in EC car emission legislation'. In D. Liefferlink, et al. (eds) *European Integration and Environmental Policy*. Oxford: Oxford University Press, pp. 150–73.

Arts, B. (1996) 'The political influence of the global environmental movement: An assessment of the UN Framework Convention on Climate Change (FCCC)'. Paper presented to the 28th International Geographical Congress, 'Environmental problems and perspectives after Rio', the Hague, 8 August.

Arts, B. (1998) *The Political Influence of Global NGOs: Case Studies on the Climate and Biodiversity Conventions*. Utrecht: International Books.

Arts, B. and Rüdig, W. (1995) 'Negotiating the "Berlin Mandate": Reflecting on the First "Conference of the Parties" to the UN Framework Convention on Climate Change'. *Environmental Politics*, 4(3) (Autumn), pp. 481–87.

Athanasiou, T. (1991) *US Politics and Global Warming*. Open Magazine Pamphlet series, no. 14. New Jersey: Westfield.

Axelrod, R. (1990) *The Evolution of Cooperation*. London: Penguin.

Bach, W. (1995) 'Coal policy and climate protection: Can the tough CO_2 reduction target be met by 2005?'. *Energy Policy*, 23(1), pp. 85–91.

Bachrach, P. (1967) *The Theory of Democratic Elitism: A Critique*. Boston: Little Brown.

Bachrach, P. (1971) *Political Elites in a Democracy*. New York: Atherton Press.

Bachrach, P. and Baratz, M. (1962) 'Two faces of power'. *American Political Science Review*, 56, pp. 947–52.

Bachrach, P. and Baratz, M. (1970) *Poverty and Power: Theory and Practice*. London, Toronto: Oxford University Press.

Bachrach, P. and Baratz, M. (1971) 'Two faces of power'. In F. G. Castles, D. J. Murray and P. C. Potter, *Decisions, Organisations and Society*, reproduced from *American Political Science Review*, 56(1962), pp. 947–52.

Bachrach, S. and Lawler, E. (1981) *Power and Politics in Organisations*. Wellington and London: Sobbey-Bass.

Bagdikian, B. (1987) *The Media Monopoly*. Boston: Beacon Press.

Baird, R. (1996) 'Stormy weather will cost more'. *Guardian*, 3 July.

Banuri, T. (1993) 'The landscape of diplomatic conflicts'. In W. Sachs (ed.) *Global Ecology*. London and New Jersey: Zed Books, pp. 49–66.

Barratt, C. (1996) *Sustainable Development Case Study*. Teaching Politics Series. York: York University.

Bate, R. (1997) 'Rio set the bar too high'. *Wall Street Journal Europe*, 25 June.

Bate, R. and Morris, W. (1994) *Apocalypse or Hot Air?*. IEA Studies on the Environment. London: IEA.

Bauer, R. A. and Gergen, K. J. (eds) (1968) *The Study of Policy Formation*. New York: Free Press; London: Macmillan.

Beavis, S. (1992) 'Electricity generators launch campaign against proposed European carbon tax'. *Guardian*, 21 May.

Becker, H. S. (1970a) 'Whose side are we on?'. In W. Filstead (ed.) *Qualitative Methodology*. Chicago: Markham, pp. 15–26.

Becker, H. (1970b) 'Problems of inference and proof in participant observation'. In W. Filstead, *Qualitative Methodology*. Chicago: Markham, pp. 189–201.

Becker, H. S and Geer, B. (1970) 'Participation observation and interviewing: A comparison'. In W. Filstead (ed.) *Qualitative Methodology*, pp. 133–52.

Beckerman, W. (1992) 'Global warming and international action: An economic perspective'. In A. Hurrell and B. Kingsbury (eds) *The International Politics of the Environment: Actors and Institutions*. Oxford: Claredon Press, pp. 253–90.

Beckerman, W. and Malkin, J. (1994) 'How much does global warming matter?'. *The Public Interest* no. 114 (Winter), pp. 4–16.

Beckerman, W. and Pasek, J. (1995) 'The equitable international allocation of tradable permits'. *Global Environmental Change*, 5(5), pp. 405–13.

Beder, S., Brown, P. and Vidal, J. (1997) 'Who killed Kyoto?'. *Guardian*, 29 October, pp. 4–5.

Bell, A. (1994) 'Climate of opinion: Public and media discourse on the environment'. *Discourse and Society*, 5(1), pp. 33–64.

Benedick, R. E. (1991a) *Ozone Diplomacy: New Directions in Safeguarding the Planet*. Cambridge MA: Harvard University Press.

Benedick, R. (1991b) 'Building on the Vienna Convention'. In WRI, *Greenhouse Warming: Negotiating a Global Regime*. Washington: WRI, pp. 9–12.

Benedick, R. (1991c) 'The diplomacy of climate change: Lessons from the Montreal ozone protocol'. *Energy Policy*, 19(2) (March), pp. 94–7.

Bennett, R. S. (1996) Executive Director of the Society for Environmental Truth, Questionnaire on the Politics of Global Warming.

Bennett, T. (1982a) 'Media, "reality", signification'. In M. Gurevitch, T. Bennett, J. Carran and J. Woollacott (eds) *Culture, Society and the Media*. London: Routledge, pp. 287–309.

Bennett, T. (1982b) 'Theories of the media, theories of society'. In M. Gurevitch, T. Bennett, J. Carran and J. Woollacott (eds) *Culture, Society and the Media*. London: Routledge, pp. 30–56.

Benton, M. and Frazier, P. J. (1991) 'The agenda-setting function of the mass media at three levels of "Information holding"'. In M. McCombs and M. Protess (eds) *Agenda-Setting: Reading on the Media, Public Opinion and Policy-making*. New Jersey: Lawrence Erlbaum, pp. 61–9.

Bergesen, H. (1989) 'The credibility of science in international resource management'. In S. Andresen and W. Ostreng (eds) *International Resource Management: The Role of Science and Politics*. New York and London: Belhaven Press, pp. 124–32.

Bergesen, H. (1991a) 'Symbol or substance? The climate policy of the EC'. *International Challenges*, 11(4), pp. 24–9.

Bergesen, H. (1991b) 'A legitimate social order in a "greenhouse world": Some basic requirements'. *International Challenges*, 11(2), pp. 21–9.

Bergesen, H. (1992) 'Empty symbols or a process that can't be reversed?: A tentative evaluation of the institutions emerging from the UN Conference on Environment and Development'. *International Challenges*, 12(3), pp. 4–14.

Bergesen, H., Grubb, M., Howcade, J. C., Jaeger, J., Lanza, A., Lobke, R., Sverdrup, L. and Tudin, A. (1994) *Implementing the European CO_2 Commitment: A Joint Policy Proposal*. London: RIIA.

Bernauer, T. (1995) 'The effect of international environmental institutions: How might we learn more'. *International Organisation*, 49(2), pp. 351–77.

Betsill, M. (1998) 'NGOs and the United Nations Framework Convention on Climate Change'. APSA paper, September.

Beuermann, C. and Jaeger, J. (1996) 'Climate change politics in Germany: How long will any double dividend last?'. In T. O'Riordan and A. Jäger (eds) *The Politics of Climate Change: A European Perspective*. London and New York: Routledge, pp. 186–228.

Blantran de Rozari, M. (1996) Government advisor to the Environment Ministry, Indonesia. Interview, Geneva, COP2, 19 July.

Blumler, J. G. and Guervitch, M. (1982) 'The political effects of mass communication'. In Gurevitch, M. et al. *Media Culture and Society*. London: Routledge, pp. 236–68.

Bodansky, D. (1995) 'The emerging climate change regime'. *Annual Review of Energy and Environment*, 20, pp. 425–61.

Boehmer-Christiansen, S. (1989) 'The role of science in the international regulation of pollution'. In S. Andresen and W. Ostreng (eds) *International Resource Management: The Role of Science and Politics*. New York and London: Belhaven Press, pp.143–68.

Boehmer-Christiansen, S. (1994a) 'Global climate policy: the limits of scientific advice: Part 1'. *Global Environmental Change*, 4(2), pp. 140–59.

Boehmer-Christiansen, S. (1994b) 'Global climate protection policy: the limits of scientific advice: Part 2'. *Global Environmental Change*, 4(3), pp. 185–200.

Boehmer-Christiansen, S. (1994c) 'A scientific agenda for climate policy?'. *Nature*, 372(1), (December), pp. 542–5.

Boehmer-Christiansen, S. (1995) 'Britain and the International Panel on Climate Change: The impacts of scientific advice on global warming part 1: Integrated policy analysis and the global dimension'. *Environmental Politics*, 4(1) (Spring), pp. 1–18.

Boehmer-Christiansen, S. (1996) 'The international research enterprise and global environmental change: Climate change policy as a research process'. In J. Vogler and M. Imber (eds) *The Environment of International Relations*. London: Routledge, pp. 171–96.

Boehmer-Christiansen, S. and Skea, J. (1991) *Acid Politics: Environmental and Energy Policies in Britain and Germany*. London: Belhaven Press.

Boehmer-Christiansen, S. and Skea, J. (1994) 'The operation and impact of the intergovernmental panel on climate change: Results of a survey of participants and users'. Paper presented at the conference on 'Governing our Environment', Copenhagen, 16–18 November.

Bostrom, A. (1994a) 'What do people know about climate change?'. *Risk Analysis*, 14(6), pp. 959–69.

Bostrom, A. (1994b) 'What do people know about climate change?'. *Risk Analysis*, 14(6) pp. 971–82.

Boulton, L. (1997) 'Japan attacked as climate deal nears'. *Financial Times*, 5 December, p. 4.

Bourillon, C. (1995) Manager of Public Affairs World Coal Institute. Interview, London, 22 June.

Bowers, C. (1993) 'Europe's motorways: The drive for mobility'. *The Ecologist*, 23(4), pp. 125–30.

Boyd-Bennett, J. O. (1982) 'Cultural dependency and the mass media'. In M. Gurevitch et al. (eds) *Culture, Society and the Media*. London: Routledge, pp. 174–97.

Boyer, B. (1997) 'Networks and means of action: NGOs and the development of Swiss climate policy'. IPSA paper, 25 July.

Boyle, S. (1998) 'Early birds and ostriches'. *Energy Economist*, June. London: Financial Times Publishing.

Boyle, S. and Ardill, J. (1989) *The Greenhouse Effect: A Practical Guide to the World's Changing Climate*. London: New English Library.

Bramble, B. and Porter, G. (1992) 'Non-governmental organisations and the making of US international environmental policy'. In A. Hurrell and B. Kingsbury (eds) *International Politics of the Environment: Actors and Institution*. Oxford: Clarendon Press, pp. 313–53.

Breitmeier, H. and Dieter Wolf, K. (1995) 'Analysing regime consequences: Conceptual outlines and environmental explorations'. In V. Rittberger (ed.) *Regime Theory and International Relations*. Oxford: Clarendon Press, pp. 339–61.

Brenton, T. (1994) *The Greening of Machiavelli: The Evolution of International Environmental Politics*. London: RIIA/Earthscan.

Breyman, S. (1993) 'Knowledge as power: Ecology movements and global environmental problems'. In R. Lipschultz and K. Conca (eds) *The State and Social Power in Global Environmental Politics*. New York: Columbia University Press, pp. 124–58.

British Roads Federation (1992, 1993, 1994) *Annual Report*. London: BRF.

Brown, C. and Schoon, N. (1992) 'Heseltine demands more intervention on global warming'. *The Independent*, 28 February, p. 17.

Brown, D. (1994) 'Farm threat in global warming is exaggerated'. *Daily Telegraph*, 28 October, p. 5.

Brown, P. (1994) environment correspondent at The *Guardian*, interview, London, 25 February.

Brown, P. (1996a) 'Change in climate of opinion'. *Guardian*, 3 July, p. 3.

Brown, P. (1996b) 'Meltdown'. *Guardian*, 6 July, pp. 30–5.

Brown, P. (1996c) *Global warming: can civilization survive?* London: Blandford Press.

Brown, P. (1997) 'Anger at deal loopholes as climate talks drag on'. *Guardian*, 11 December, p. 3.

Browne, A. (1998) 'Industry green about the gills over energy tax'. *Observer*, 24 May, p. 4.

Bull, H. (1977) *The Anarchical Society: A Study of Order in World Politics*. London: Macmillan.

Bundesverband der Deutschen Industrie (BDI) (1990) *Precaution against Possible Climate Change: A Stocktaking by German Industry*. BDI, January.

Burgess, J. and Harrison, C. M. (1993) 'The circulation of claims in the cultural politics of environmental change'. In A. Hansen (ed.) *The Mass Media and Environmental Issues*. Leicester: Leicester University Press, pp. 198–222.

Burgess, J., Harrison, C. and Maitney, P. (1991) 'Contested meanings: The consumption of news about nature conservation'. *Media, Culture and Society*, 13, pp. 499–520.

Business and Industry (1996) Statement to the second meeting of the Conference of the Parties to the UNFCCC, Geneva, 12 July.

Cain, M. (1983) 'Carbon dioxide and the Climate: Monitoring and a search for understanding'. In D. Kay and H. K. Jacobson, *Environmental Protection: The International Dimension*. New Jersey: Allonheld Osmun, pp. 75–98.

Caldwell, L. K. (1988) 'Beyond environmental diplomacy: the changing institutional structure of international cooperation'. In J. Carroll (ed.) *International Environmental Diplomacy*. Cambridge: Cambridge University Press, pp. 13–28.

Caldwell, L. K. (1990) *International Environmental Policy: Emergence and Dimensions*, 2nd edn. DURHAM, NC: Duke University Press.

Capoor, K. (1996) Environmental Defense Fund (United States), Interview, Geneva, COP2, 15 July.

Carlo, L. and Garcia, M. (1993) 'Strategies and perspectives towards climate change in Spain'. In P. Vellinga and M. Grubb (eds) *Climate Change Policy in the European Community*. Report of October 1992 Workshop. London: RIIA, pp. 39–48.

Carr, S. and Mpende, R. (1996) 'Does the definition of the issue matter? NGO influence and the international convention to combat desertification in Africa'. In D. Potter (ed.) *NGOs and Environmental Policies: Asia and Africa*. London and Oregon: Frank Cass, pp. 143–58.

Carroll, J. (1988) *International Environmental Diplomacy*. Cambridge: Cambridge University Press.

Cavender, J. and Jaeger, J. (1993) 'The History of Germany's Response to Climate Change'. *International Environmental Affairs*, 5(1) (Winter), pp. 3–18.

Cawson, A. (1992) 'Interests, groups and public-policy-making: The case of the European consumer electronics industry'. In J. Greenwood, J. R. Grote and K. Ronit (eds) *Organised Interests and the European Community*. London: Sage.

Chapman, G. (1992) 'TV: The world next door?'. *Intermedia*, 20(1), pp. 23–8.

Chapman, G. (1997) 'Environmentalism, the mass media and the global silent majority'. *Global Environmental Change Programme Briefings*, no. 15 (November).

Chapman, G., Keval, K., Fraser, C. and Gaber, I. (1997) *Environmentalism and the Mass Media: The North-South Divide*. London: Routledge.

Chatterjee, P. and Finger, M. (1994) *The Earth Brokers: Power, Politics and World Development*. London: Routledge.

Chomsky, N. (1989) *Necessary Illusions: Thought Control in Democratic Societies*. London: Pluto.

Clegg, S. R. (1989) *Frameworks of Power*. London, Newbury Park and New Delhi: Sage.

Clegg, S. R. and Dunkereley, D. (1980) *Organisation, Class and Control*. London: Routledge and Kegan Paul.

Climate Action Network (1995) *Independent NGO Evaluations of National Plans for Climate Change Mitigation: OECD Countries and Central and Eastern Europe*, 1st and 3rd reviews. Brussels: CNE.

Climate Change Bulletin (1995a) 2nd quarter, issue 7.

Climate Change Bulletin (1995b) 4th quarter, issue 9.

Climate Network Africa (1993) 'Study of the IPCC Greenhouse Gas Inventory Methodology Applied to Land Use and Forestry Changes in Kenya'. Nairobi: CNA.

Climate Network Europe (1994) *Joint Implementation from a European NGO Perspective*, July. Brussels: CNE.

Climate Network Europe/Climate Action Network-US (1996) *Independent NGO Evaluation of National Plans for Climate Change Mitigation*, 4th (interim) review. Brussels: CNE.

Cline, W. (1993) 'Costs and benefits of greenhouse abatement: A guide to policy analysis', international conference on the economics of climate change. Paris: OECD.

Clover, C. (1997) 'Pollution treaty in the balance after "rebellion"'. *Daily Telegraph*, 11 December, p. 20.

Clover, C. (1998) 'Clinton on course for a clash with Europe over global warming'. *Daily Telegraph*, 15 May, p. 14.

Cohen, B. (1963) *The Press and Foreign Policy*. New Jersey: Princeton University Press.

Colat, P. (1996) Department of the Environment (Canada), Interview, Geneva, COP2, 16 July.

Colglazier, E.W. (1991) 'Scientific uncertainties, public policy, and global warming: How sure is sure enough?'. *Policy Studies Journal*, 19(2) (Spring), pp. 61–72.

Collie, L. (1993) 'Business Lobbying in the European Community: The Union of Industrial and Employers' Confederations of Europe'. In S. Mazzey and J. Richardson (eds) *Lobbying in the EC*. Oxford: Oxford University Press, pp. 213–27.

Collier, U. (1993) 'Subsidiarity and Climate Change Policy: An Excuse for Inaction in the European Community?' Paper presented at the Conference on 'Perspectives on the Environment 2: Research and Action' University of Sheffield, 13–14 September.

Collier, U. and Löfstedt, R. E. (eds) (1997) *Cases in Climatic Change Policy: Political Reality in the European Union*. London: Earthscan.

Conca, K. (1993) 'The environment and the deep structure of world politics'. In R. Lipschultz and K. Conca (eds) *The State and Social Power in Global Environmental Politics*. New York: Columbia University Press, pp. 306–27.

Conca, K. (1995) 'Greening the UN: Environmental organisations and the UN system'. *Third World Quarterly*, 16(3), pp. 103–19.

Confederation of British Industry (1993) 'Climate change: Our National Programme for CO_2 Emissions'. Memorandum, April. London: CBI.

Congress of the US (1996) Letter to Warren Christopher, US Secretary of State, 10 July.

Corner House, The (1997) 'Climate and equity after Kyoto'. Briefing 3, December.

Corner, J. (1991) Editorial, *Media, Culture and Society*, 13(4) (October), pp. 435–41.

Corner, J. and Richardson, K. (1993) 'Environmental communication and the contingency of meaning: A research note'. In A. Hansen (ed.) *The Mass Media and Environmental Issues*. Leicester: Leicester University Press, pp. 222–34.

Cottle, S. (1993) 'Mediating the environment: Modalities of TV news'. In A. Hansen (ed.) *The Mass Media and Environmental Issues*. Leicester: Leicester University Press, pp. 107–34.

Cowe, R. (1998) 'CBI attacks plan for energy tax'. *Guardian*, 5 August, p. 16.

Cox, R. (1981) 'Social Forces, States and World Orders: Beyond International Relations theory'. *Millennium*, 10(2), pp. 126–55.

Cox, R. (1987) *Production, Power and World Order: Social Forces in the Making of History*. New York: Columbia University Press.

Cox, R. (1994) 'Social Forces, States and World Orders: Beyond International Relations Theory'. In F. Kratochwil and E. Mansfield, International Organisation: A Reader. New York: HarperCollins, pp. 343–64.

Cox, R. W. and Jacobsen, H. K. (1973) *The Anatomy of Influence: Decision-making in International Organizations*. New Haven: Yale University Press.

Cozijnsen, J. (1996) Ministry of Housing, Spatial Planning and the Environment (Netherlands), Interview, Geneva, COP2, 15 July.

Cracknell, J. (1993) 'Issue arenas, pressure groups and environmental agendas'. In A. Hansen (ed.) *The Mass Media and Environmental Issues*. Leicester: Leicester University Press, pp. 3–22.

Crandall, C. (undated) 'International scientific panel completes analysis of IPCC reports: Finds no support for greenhouse warming or global climate treaty', Science and Environmental Policy Project, pp. 1–3.

Crenson, M. (1971) *The UnPolitics of Air Pollution*. London and Baltimore: Johns Hopkins Press.

Curran, J. (1982) 'Communications power and social order'. In M. Gurevitch et al. (eds) *Culture, Society and the Media*. London: Routledge, pp. 202–36.

Curran, J. (1986) 'The impact of advertising on the British mass media'. In R. Collins, J. Curran, N. Garnham, P. Scannell, P. Schlesinger and C. Sparks (eds) *Media Culture and Society: A Critical Reader*. London: Sage, pp. 309–36.

Curran, J. (1988) *Power without Responsibility: The Press and Broadcasting in Britain*, 3rd edn., London: Routledge.

Curran, J., Gurevitch, M. and Woollacott, J. (1977) *Mass Communication and Society*. London: Edward Arnold.

Curran, J. Gurevitch, M. and Woollacott, J. (1982) 'The study of the media: theoretical approaches'. In M. Gurevitch et al. (eds) *Culture, Society and the Media*. London: Routledge, pp. 11–30.

Dahl, A. (1994) 'Energy and environment in Western Europe'. *International Challenges*, **14**(1), pp. 23–32.

Dahl, R. (1957) *A Preface to Democratic Theory*. Chicago: Chicago University Press.

Dahl, R. (1961) *Who Governs?: Democracy and Power in an American City*. New Haven and London: Yale University Press.

Dahl, R. (1963) *Modern Political Analysis*. Englewood Cliffs: Prentice Hall.

Dahlgren, P. (1992) 'What's the meaning of this? Viewers' plural sense-making of TV news'. In P. Scannell, P. Schlesinger and C. Sparks (eds) *Culture and Power*. London: Sage, pp. 201–18.

Daily Express (1991) 'Global warming is a load of hot air', 8 December, p. 4.

Daily Telegraph (1991) 'The lie of global warming', 24 April, p. 2.

Daily Telegraph (1992) '10pc rise in fuel prices plan to cut carbon pollution', 15 December, p. 3.

Daily Telegraph (1994a) 'Carbon tax no use say greenhouse sceptics', 17 March, p. 1.

Daily Telegraph (1994b) 'Sun main culprit of global warming', 7 April, p. 2.

Dasgupta, C. (1994) 'The climate change negotiations'. In I. M. Mintzer and J. A. Leonard (eds) *Negotiating Climate Change*. Cambridge: Cambridge University Press, pp. 129–49.

Davies, J. (1994) Environment correspondent at the *Daily Express*, interview, 2 March.

Davis, G. (1989) 'Global warming: The role of energy efficient technologies'. Shell Selected Papers, October.

Dawson, B. (1992) 'EPA signals unchanged "greenhouse gas" policy'. *Houston Chronicle*, 15 February, p. 18.

Day, D. (1989) *The Ecowars: A Layman's Guide to the Ecology Movement*. London: Harrop.

De Vaus, D. A. (1991) *Surveys in Social Research*, 3rd edn. London: UCL Press.

Djoghlaf, A. (1994) 'The beginning of an international climate law'. In I. Mintzer and J. A. Leonard (eds) *Negotiating Climate Change*. Cambridge: Cambridge University Press, pp. 97–113.

Doherty, A. and Hoedeman, O. (1994) 'Misshaping Europe: The European Roundtable of Industrialists'. *The Ecologist*, **24**(4) (July/August), pp. 135–41.

Domini, A. (1996) 'The bureaucracy and the free spirits: Stagnation and innovation in the relationship between the UN and NGOs'. In T. G. Weiss and L. Gordeneker (eds) *NGOs, the UN and Global Governance*. London and Boulder: Lynne Riener, pp. 83–101.

Donovan, P., Whitebloom, S. and Hencke, D. (1994) 'Lobby firm loses blue chip clients'. *Guardian*, 22 October, p. 19.

Doran, P. (1993) 'The Earth Summit: Ecology as spectacle'. *Paradigms*, **7**(1) (Summer), pp. 55–65.

Douthwaite, R. (1994) 'Commons and goings'. *Guardian*, 1 July, p. 27.

Douthwaite, R. (1995) 'Who says that life is cheap'. *Guardian*, 1 November, p. 18.

Dowdeswell, E. and Kinley, R. (1994) 'Constructive damage to the status quo'. In I. M. Mintzer and J. A. Leonard (eds) *Negotiating Climate Change*. Cambridge: Cambridge University Press, pp. 113–29.

Dowding, K. M. (1991) *Rational Choice and Political Power*. Aldershot: Edward Elgar.

Downs, A. (1972) 'Up and down with ecology: The issue attention cycle'. *Public Interest*, **28**, pp. 38–50.

Downs, A. (1991) 'Up and down with ecology: The issue-attention cycle'. In M. McCombs and M. Protess (eds) *Agenda-Setting: Readings on Media, Public Opinion and Policy-making*. New Jersey: Lawrence Erlbaum, pp. 7–33.

Doyle, L. (1992a) 'US sold out to coal and oil industry'. *The Independent*, 25 February, p. 6.

Doyle, L. (1992b) 'US cools towards global warming'. *The Independent*, 26 February, p. 14.

Dreyfus, H. and Rabinow, P. (1983) *Michel Foucault: Beyond Structuralism and Hermeneutics*, 2nd edn. Chicago: University of Chicago Press.

Dubash, N. K. and Oppenheimer, M. (1992) 'Modifying the Mandate of Existing Institutions: NGOs'. In I. Mintzer (ed) *Confronting Climate Change*. Cambridge: Cambridge University Press, pp. 265–81.

Dunne, N. (1992) 'Bush may be warming to Rio's Earth Summit'. *The Independent*, 22 March, p. 13.

Dunwoody, S. and Griffin, R. J. (1993) 'Journalistic strategies for reporting long-term environmental issues: a case study of three Superfund sites'. In A. Hansen (ed.) *The Mass Media and Environmental Issues*. Leicester: Leicester University Press, pp. 22–51.

Dykins, P. (1994) Global Atmosphere Division, DoE, Marsham Street, London, interview, 25 February.

Easterbrook, G. (1992) 'Has Environmentalism Blown it?'. *The New Republic*, 6 July, pp. 23–5.

Eberlie, R. (1993) 'The confederation of British industry and policy-making in the European Community'. In S. Mazzey and J. Richardson (eds) *Lobbying in the EC*. Oxford: Oxford University Press, pp. 201–12.

Eccleston, B. (1996) 'Does North–South collaboration enhance NGO influence on deforestation policies in Malaysia and Indonesia?'. In D. Potter (ed.) *NGOs and Environmental Policies: Asia and Africa*. London and Oregon: Frank Cass, pp. 66–85.

Eccleston, B. (1996a) 'NGOs and competing representations of deforestation as an environmental issue in Malaysia'. In D. Potter (ed.) *NGOs and Environmental Policies: Asia and Africa*. London and Oregon: Frank Cass, pp. 116–38.

EC Energy Monthly (1996) 'Eight ministers call for progress on stalled CO_2 energy tax', issue 86 (February).

ECInform-Energy (1995) issue 33 (December).

ECOAL 2 (undated) 'Coal technology will overcome environmental objections'. London: World Coal Institute, p. 2.

ECOAL 4 (undated) 'Carbon tax would send coal and gas prices soaring', 'New coal technologies will lower CO_2 emissions'. London: World Coal Institute, p. 3.

ECOAL 7 (undated) 'Experts express uncertainty over global warming'. London: World Coal Institute, p. 1.

ECOAL 8 (undated) 'Global warming theory may be falling apart', London: World Coal Institute, p. 2.

ECOAL 15 (1995) 'EU says "no" to carbon tax', January. London: World Coal Institute, p. 4.

ECOAL 18 (1996) 'Bombshell lands in the climate change arena', July. London: World Coal Institute, p. 5.

Economist (1987) 'Castor oil or Camelot?', 5 December, p. 27.

Economist (1990) 'Warm world cool heads', 27 October, p. 13.

Economist (1992a) 'Europe's Industries Play Dirty', 9 May, pp. 91–2.

Economist (1992b) 'Taxing Carbon', 9 May, pp. 18–19.

Economist (1995) 'Stay cool', 1 April, p. 11.

Economist (1998) 'Global warming: In flux', 11–17 April, p. 97.

ECSC (European Coal and Steel Community) (1992) 'Resolution of the consultative committee of the European Coal and Steel Community', Brussels: ECSC. 3 April.

Eden, S. (1994) 'Using sustainable development: The business case'. *Global Environmental Change*, 4(2), pp. 160–7.

Edwards, D. (1997) 'Hot air: Global warming and the political economy of threats'. *The Ecologist*, 27(1) (January/February), pp. 2–4.

Edwards, D. (1998) 'Can we learn the truth about the environment from the media?'. *The Ecologist*, 28(1) (January/February), pp. 18–22.

Eikeland, P. (1993a) 'The shaping of US energy policy'. *International Challenges*, 13(4), pp. 3–16.

Eikeland, P. (1993b) 'US energy policy at the cross-roads?'. *Energy Policy*, 21(10) (October), pp. 987–98.
Eikeland, P. (1993c) 'US environmental NGOs in the greenhouse – towards changed strategies?'. Energy, Environment and Development Working Paper, 1993/5.
Einsiedel, E. and Coughlan, E. (1993) 'The Canadian press and the environment: reconstructing a social reality'. In A. Hansen (ed.) *The Mass Media and Environmental Issues*. Leicester: Leicester University Press, pp. 134–50.
El-Ashry, M. (1995) Chief Executive officer of GEF, Letter to author, 17 April.
Eldridge, J. (1993) *Getting the Message: News, Truth and Power*. London: Routledge.
Eleri, E. O. (1994) 'Africa's decline and greenhouse politics'. *International Environmental Affairs*, 6(2) (Spring), pp. 133–48.
Elliott, L. (1992) 'Non-governmental organisations in international environmental politics: practice and theory'. BISA paper, University of Swansea, December.
Elliott, L. (1994) 'Spurious followers of green fashion'. *Guardian*, 24 January, p. 13.
Elliott, P. (1977) 'Media organisations and occupations: An overview'. In J. Curran, M. Gurevitch and J. Woollacott (eds) *Mass Communication and Society*. London: Edward Arnold, pp. 142–74.
ENDS Report (1992a) 'The great carbon tax debate', no. 204 (January).
ENDS Report (1992b) 'Individual winners and losers in the carbon tax game', no. 207 (April).
ENDS Report (1992c) 'Carbon-friendly industrialists', no. 207 (April).
ENDS Report (1992d) 'Temporary set-back for EC's strategy on CO_2', no. 208 (May).
ENDS Report (1992e) 'DTI consults industry as carbon tax debate advances', no. 211 (August).
ENDS Report (1992f) 'Deadlock over CO_2 controls on cars', no. 215 (December).
ENDS Report (1993a) 'HMIP disappoints with power station authorisations', no. 219 (April).
ENDS Report (1993b) 'German industries follow Dutch in volunteering CO_2 reductions', no. 244 (May).
ENDS Report (1993c) 'UK to block EC carbon tax', no. 220 (May).
ENDS Report (1994a) 'Science and nonsense in the global warming debate', no. 233 (June).
ENDS Report (1994b) 'Growing tension over global warming review', no. 236 (September).
Engler, R. (1967) *The Politics of Oil: A Study of Private Power and Democratic Directions*. New York: Phoenix.
Environment Bulletin (1995) 71(3).
Environment Business (1991a) 'Euro-Carbon tax concept stumbles again', 31 July, p. 3.
Environment Business (1991b) 'EC steel lobby attacks proposed energy tax', 25 September, p. 2.
Environment Business (1991c) 'Industry wins first EC energy tax round', 9 October, p. 5.
Environment Business (1991d) 'Wind lobby calls for a major expansion', 23 October, p. 4.
Environment Business (1991e) 'Dutch companies fight doubling of energy tax', 23 October, p. 8.
Environment Business (1992a) 'Pressure builds against CO_2 tax', 6 May, p. 4.
Environment Business (1992b) 'Dutch postpone decision on eco-energy levies', 3 March, p. 6.
Environment Business (1992c) 'Dutch in crisis over eco-levies', 12 September, p. 7.
Environment Business (1992d) 'Transport sector hampers German CO_2 cuts effort', 7 October, p. 4.
Environment Business (1993a) 'Energy efficiency laws could backfire', 10 February, p. 5.
Environment Business (1993b) 'Dutch employers oppose unilateral energy levy', 24 April, p. 1.
Environment Digest (1994) 'UK may miss energy saving targets', no. 80 (February), p. 10.
Environment Digest (1995) 'EU carbon tax guidelines', no. 5–6, p. 8.
Environment Digest (1996) 'Climate report criticised', no. 4 (April–May), p. 9.
Eurelectric (1991a) 'Proposal of the European electric supply industry for an effective EC strategy for greenhouse gas control', 24 June. Brussels: Eurelectric.
Eurelectric (1991b) 'Eurelectric's position on the Community strategy to limit carbon dioxide emissions to improve energy efficiency', 5 November. Brussels: Eurelectric.
European Business Council for a Sustainable Energy Future (1996) Newsletter, September.

European Coal and Steel Community (1992) 'Resolution of the consultative committee of the European Coal and Steel Community' (92/C 127/02), 3 April. Brussels: ECSC.

European Roundtable of Industrialists (1994) 'The climate change debate: Seven principles for practical policies'. Working paper prepared by the ERT Environment Watchdog, December. Brussels: ERT.

European Roundtable of Industrialists (1997) 'Climate change: An ERT report on positive action'. ERT report for Kyoto. Brussels: ERT.

Evans, P., Rueschemeyer, D. and Skocpol, T. (eds) (1985) *Bringing the State Back In.* Cambridge, New York and Melbourne: Cambridge University Press.

Evening Standard (1992a) 'Warming may bring malaria to Britain', 12 February, p. 19.

Evening Standard (1992b) 'The world faces series of disasters', 14 February, p. 16.

Falk, R. (1971) *This Endangered Planet: Prospects and Proposals for Human Survival.* New York: Vintage Books.

Farley, M. (1997) 'Showdown at global warming summit'. *Los Angeles Times*, 8 December.

Faulkner, H. (1994) 'Some comments on the INC process'. In I. Mintzer and J. A. Leonard (eds) *Negotiating Climate Change.* Cambridge: Cambridge University Press, pp. 229–39.

Fay, K. and Stirpe, D. (1995) 'Final report from the first meeting of the Conference of the Parties to the Framework Convention on Climate Change'. Memo to ICCP, 17 April.

Fermann, G. (ed.) (1997) *International Politics of Climate Change: Key Issues and Critical Actors.* Oslo: Scandinavian University Press.

Financial Times (1993) 'Energy industry warns of heavy cost', 19 February, p. 17.

Finer, S. (1971) 'The political power of private capital'. In F. G. Castles, D. J. Murray and D. C. Potter (eds) *Decisions, Organisations and Society.* Harmondsworth: Penguin Books.

Finger, M. (1993) 'Politics of the UNCED process'. In W. Sachs (ed.) *Global Ecology.* London and New Jersey: Zed Books, pp. 36–48.

Finger, M. (1994) 'NGOs and transformation: Beyond social movement theory'. In T. Princen and M. Finger *Environmental NGOs and World Politics.* London: Routledge, pp. 48–69.

Fischer, K. and Schot, J. (eds) (1993) *Environmental Strategies for Industry: International Perspectives on Research Needs and Policy Implications.* Washington: Island Press.

Fish, A. L. Jr. and South, D. W. (1994) 'Industrialised countries and greenhouse gas emissions'. *International Environmental Affairs*, 6(1) (Winter), pp. 14–43.

Fisher, D. (1995) 'The emergence of the environmental movement in Eastern Europe and its role in the revolutions of 1989'. In K. Conca, M. Alberty and G. Dubelko (eds) *Green Planet Blues: Environmental Politics from Stockholm to Rio.* Colorado and Oxford: Westview, pp. 107–15.

Fleagle, R. (1992) 'The US government response to global change: analysis and appraisal'. *Climatic Change*, 20, pp. 57–81.

Forbes (1989) 'The global warming panic', 13 December, p. 1.

Forsyth, T. (1998) 'Renewable energy investment and technology transfer in Asia'. Workshop report, Briefing Paper No. 51, October, Chatham House, London.

Forum Europe March (1992) *Carbon Tax Survey.* Brussels: Forum Europe.

Foucault, M. (1977) *Discipline and Punish: The Birth of the Modern Prison.* London: Allen Lane.

Foucault, M. (1980) 'Truth and power'. In M. Foucault, *Power/Knowledge: Selected interviews and other writings 1972–1977.* Brighton: Harvester Press.

Foucault, M. (1983) *The History of Sexuality: Three volumes*, 2nd edn. London: Lane.

Foucault, M. (1988) *Politics, Philosophy, Culture: Interviews and Other Writings, 1977–1984*, edited by L. Kvitzman. London: Routledge.

Freidrich, I. (1937) *Constitutional Government and Democracy.* New York: Gipp Press.

Friends of the Earth-UK (1994) 'The Climate Resolution: How local authorities can join forces to help take the heat off the planet', January. London: FoE-UK.

Galbraith, J. K. (1984) *The Anatomy of Power.* London: Hamish Hamilton.

Gallon Environment Letter (1998) 2(16) (30 June).

Gan, L. (1993) 'The making of the Global Environmental Facility'. *Global Environmental Change*, 3(3) (September), pp. 256–73.

Gandy, O. (1991) 'Beyond agenda-setting'. In M. McCombs and M. Protess (eds) *Agenda-Setting: Readings on Media, Public Opinion and Policy-making*. New Jersey: Lawrence Erlbaum, pp. 264–73.

Garnham, N. (1986) 'The media and the public sphere'. *Intermedia*, 14(1) (January), pp. 28–33.

Ghazi, P. (1992) 'Rising seas may drown Hong Kong, Tokyo, Rio'. *Observer*, 29 March, p. 15.

Ghazi, P. (1997) 'Climate deal undermined by sceptics'. *Observer*, 9 March.

Gill, S. and Law, D. (1988) *The Global Political Economy: Perspectives, Problems and Policies*. Herts: Harvester Wheatsheaf.

Gill, S. (1994) 'Knowledge, politics and the neo-liberal political economy'. In R. Stubbs and G. Underhill (eds) *Political Economy and the Changing Global Order*. Basingstoke: Macmillan, pp. 75–89.

Gilpin, R. (1987) *The Political Economy of International Relations*. Princeton NJ: Princeton University Press.

Global Climate Coalition (undated) 'Potential climate change: Issues and options'.

Global Climate Coalition (1992) 'Global Climate Coalition endorses administration proposal to create technology fund', statement, 27 February.

Global Climate Coalition (undated) briefings distributed at INC meetings: 'Greenhouse controls costly, ineffective', 'Current scientific understanding of global warming does not support drastic actions', 'Carbon dioxide emissions are only part of the greenhouse effect'.

Global Climate Coalition (1995a) Position statement on First Meeting of the Conference of the Parties, Washington, US.

Global Climate Coalition (1995b) Press briefings of 7 April and 12 April, 'Developing countries escape climate treaty negotiations with no new obligations'.

Global Climate Coalition (1995c) *Climate Watch* (the bulletin of the GCC), 3(5).

Global Climate Coalition (1996) 'Global climate coalition: An overview'. Backgrounder, 23 February.

Global Commons Institute (1994) 'The Unequal Use of the Global Commons'. Paper for the IPCC Working Group, 3 July. London: GCI.

Global Commons Institute (1996) 'IPCC second assessment report press launch'. Press release, 26 June. London: GCI.

Global Environmental Facility (GEF) (1995) 'Joint Summary of the Chairs', GEF Council meeting, 3–5 May and 22–24 February. Washington: GEF.

Global Market Review (1998) 'Plan of Inaction', 26 December, p. 26.

Goldemberg, J. (1994) 'The road to Rio'. In I. M. Mintzer and J. A. Leonard (eds) *Negotiating Climate Change*. Cambridge: Cambridge University Press, pp. 175–87.

Golding, P. (1981) 'The missing dimensions: News media and the management of social change'. In E. Katz and T. Szecsko (eds) *Mass Media and Social Change*. London: Sage, pp. 63–83.

Goldmann, K and Sjöstedt, G. (eds) (1979) *Power, Capabilities, Interdependence: Problems in the Study of International Influence*. London: Sage.

Gordenker, L. and Weiss, T. G. (eds) (1995) *NGOs, the United Nations and Global Governance*. London and Boulder: Lynne Riener.

Gordon, C. (1980) *Power/Knowledge: Selected Interviews and Other Writings by Michel Foucault 1972–77*. New York: Pantheon Books.

Gowa, J. (1986) 'Anarchy, egoism, and third images: The Evolution of Co-operation and international relations'. *International Organisation*, 40(1) (Winter), pp. 167–86.

Grant, W. (1993a) 'Transnational Companies and Environmental Policy-Making: The Trend of Globalisation' In J. Liefferlink, P. Lowe and A.P. Mol (eds) (1993) *European Integration and Environmental Policy*. London: Belhaven, pp. 59–74.

Grant, W. (1993b) 'Pressure groups and the European Community: An overview'. In S. Mazzey and J. Richardson (eds) *Lobbying in the EC*. Oxford: Oxford University Press, pp. 27–45.

Grant, W. (1993c) *Business and Politics in Britain*, 2nd edn. Basingstoke: MacMillan.

Grant, W., Matthews, D. and Newell, P. (2000) *The Effectiveness of EU Environmental Policy*. Basingstoke: Macmillan.

Gray, C. B. and Rivkin, D. B. (1991) 'A "no-regrets" environmental policy'. *Foreign Policy*, no. 83 (Summer), pp. 47–65.

Green, J. and Sands, D. (1992) 'Establishing an international system for trading pollution rights'. *International Environmental Reporter*, 12 February.

Greene, O. (1991) 'Building a global warming regime: the significance of verification and confidence building measures'. Paper presented at the conference on 'International arrangements for reaching environmental goals', University of Strathclyde, 10–12 September.

Greene, O. (1996) 'Environmental regimes: Effectiveness and implementation review'. In J. Vogler and M. Imber (eds) *The Environment and International Relations*. London: Routledge, pp. 196–215.

Greene, O. and Salt, J. (1994) 'Developing an Effective Climate Change Regime'. Paper presented at the BISA Conference, York University, 19–21 December.

Greenpeace Business (1993a) 'Insurance industry taking climate change seriously', April. London: Greenpeace UK.

Greenpeace Business (1993b) 'Clean energy industry calls for CO_2 tax', June. London: Greenpeace UK.

Greenpeace Business (1993c) 'Insurance industry asks Greenpeace for assistance', August. London: Greenpeace UK.

Greenpeace Business (1994a) 'CO_2 court case: EC funded power stations in jeopardy', February. London: Greenpeace UK.

Greenpeace Business (1994b) 'Supermarket greenfreeze', October. London: Greenpeace UK.

Greenpeace Business (1995) 'Insurers confront carbon club at Berlin climate summit', April. London: Greenpeace UK.

Greenpeace International (undated) 'Fossil Fuels in a Changing Climate', summary briefing. Amsterdam: Greenpeace International.

Greenpeace International (1991a) *Global Warming: Considerations for Africa*. Submission to the Pan-African Conference on Environment and Development, Banuko, Mali January. Amsterdam: Greenpeace International.

Greenpeace International (1991b) 'Climate scientists fear effects of underestimating global warming poll shows'. Press release, 20 February. Amsterdam: Greenpeace International.

Greenpeace International (1993a) 'Energy without oil: The technical and economic feasibility of phasing out oil', January. Amsterdam: Greenpeace International.

Greenpeace International (1993b) 'Fossil fuels in a changing climate: How to protect the world's climate by ending the use of coal, oil and gas', April. Amsterdam: Greenpeace International.

Greenpeace International (1993c) 'Emerging impacts of climate change: How lucky do you feel?', April. Amsterdam: Greenpeace International.

Greenpeace International (1994a) *The Climate Time Bomb*. Amsterdam: Greenpeace International.

Greenpeace International (1994b) 'Climate change and the private financial institutions: A case of unappreciated risks and unrecognised opportunities', November. Amsterdam: Greenpeace International.

Greenpeace International (1997) *Oiling the machine: Fossil fuel dollars in the US political process*, November. Amsterdam: Greenpeace International.

Greenpeace International (1998) *The oil industry and climate change: A Greenpeace briefing*. Amsterdam: Greenpeace International.

Greenpeace (UK) (1994) *Potential Impacts of Climate Change on Health in the UK*, June. London: Greenpeace.

Greenpeace USA (1998) 'Fuelling global warming: Federal subsidies to oil in the United States'. Produced by Industrial Economics Inc. Washington: Greenpeace USA.

Greenslade, R. (1996) 'Why global warming isn't hot off the press'. *Observer*, 21 July.

Greenwood, J. and Ronit, K. (1992) 'Conclusions: Evolving patterns of organising interests in the European Community'. In J. Greenwood, et al. (eds) *Organised Interests and the EC*. London: Sage, pp. 238–52.

Greenwood, J. and Ronit, K. (1994) 'Interest groups in the European Community: Newly emerging dynamics and forms'. *West European Politics*, 17(1) (January), pp. 31–52.

Greenwood, J., Grote J. and Ronit, K. (eds) (1992) *Organised Interests and the EC*. London: Sage.

Gregory, K. and Harrison, J. (1995) *Climate Change and the work of the IPCC*. London: World Coal Institute.

Greico, J. (1988) 'Anarchy and the limits of co-operation: A Realist critique of the newest liberal institutionalism'. *International Organisation*, 42(3) (Summer), pp. 485–507.

Gribben, R. (1992) 'EC to review plans for carbon tax'. *Daily Telegraph*, 27 May, p. 12.

Gribbin, J. (1990) 'An assault on the climate consensus'. *New Scientist*, 15 December, p. 27.

Grove, J. W. (1971) 'The collective organisation of industry'. In F. G. Castles, D. J. Murray and D. C. Potter (eds) *Decisions, Organisations and Society*. Harmondsworth: Penguin Books.

Grubb, M. (1990) 'The greenhouse effect: negotiating targets'. *International Affairs*, 66(1) (July), pp. 67–89.

Grubb, M. (1991) *Energy Policies and the Greenhouse Effect Volume One: Policy Appraisal*. London: RIIA/Dartmouth.

Grubb, M. (1992) 'The heat is on'. *Times Higher Education Supplement*, 5 June, p. 14.

Grubb, M. (1995a) 'The Berlin conference: Outcome and implications'. Briefing Paper no. 21, June. London: RIIA.

Grubb, M. (1995b) 'Seeking fair weather: Ethics and the international debate on climate change'. *International Affairs*, 71(3) (July), pp. 463–96.

Grubb, M. and Anderson, D. (eds) (1995) *The Emerging International Regime for Climate Change*. London: RIIA.

Grubb, M. with Brack, D., Vrolijk, C. and Lanchberry, J. (1999) The Kyoto Protocol: A Guide and Assessment. London: RIIA/Earthscan.

Grubb, M., Brackley, P., Ledic, M., Mathur, A., Rayner, S., Russell, J. and Tanabe, A. (1991) *Energy Policies and the Greenhouse Effect. Volume Two: Country Studies and Technical Options*. London: RIIA/Dartmouth.

Grubb, M., Koch, M., Munson, A., Sullivan, F. and Thomson, K. (1993) *The Earth Summit Agreements: A Guide and Assessment*. London: Earthscan.

Grubb, M., Sebenius, J., Magalhaes, A. and Subak, S. (1992) 'Sharing the burden'. In I. Mintzer (ed.) *Confronting Climate Change: Risks, Implications and Responses*. Cambridge: Cambridge University Press, pp. 305–23.

Grubb, M. and Steen, N. (1991) *Pledge and Review Processes: Possible Components of a Climate Convention*. Report of a workshop held on, 2 August. London: RIIA.

Guardian (1992a) 'EC plans carbon fuel tax', 18 February, p. 5.

Guardian (1992b) 'Too much gas and too little action', 14 May, p. 16.

Guardian (1992c) 'Electricity generators launch campaign against proposed European carbon tax', 21 May, p. 13.

Gupta, J., Junne, G. and Wurff, R. (1993) *Determinants of Regime Formation*. Working Paper 1 for study on 'International policies to Address the greenhouse effect'. Dept. of International Relations and Public International Law, University of Amsterdam.

Gupta, S. and Pachauri, R. K. (eds) (1989) *Global Warming and Climate Change: Perspectives from Developing Countries*. New Delhi: TERI.

Gurevitch, M. and Blumer, J. (1977) 'Linkages between the mass media and politics: A model for the analysis of political communications systems'. In J. Curran, M. Gurevitch and J. Woollacott (eds) *Mass Communication and Society*. London: Edward Arnold, pp. 270–91.

Gurevitch, M., Bennett, J., Curran, J. and Woollacott, J. (eds) (1982) *Culture, Society and the Media*. London: Routledge.

Gutting, G. (ed.) (1994) *The Cambridge Companion to Foucault*. Cambridge: Cambridge University Press.

Haas, E. (1980) 'Why collaborate? Issue linkage and international regimes'. *World Politics*, **32**(3) (April), pp. 357–405.

Haas, E. (1990) *When Knowledge is Power*. Berkeley: University of California Press.

Haas, P. (1989) 'Do regimes matter? Epistemic communities and Mediterranean pollution control'. *International Organisation*, **43**(3) (Summer), pp. 377–403.

Haas, P. (1990a) *Saving the Mediterranean: The Politics of International Environmental Cooperation*. New York: Columbia University Press.

Haas, P. (1990b) 'Obtaining international environmental protection through epistemic consensus'. *Millennium*, **19**(3), pp. 347–63.

Haas, P. (1992) 'Epistemic communities and international policy coordination'. *International Organisation*, **46**(1) (Winter), pp. 1–35.

Haas, P. (1995) 'Epistemic communities and the dynamics of international environmental cooperation'. In V. Rittberger (ed.) *Regime Theory and International Relations*. Oxford: Clarendon Press, pp. 168–202.

Haas, P., Keohane, R. and Levy, M. (1993) *Institutions for the Earth: Sources of Effective Environmental Protection*. Cambridge, MA: MIT Press.

Hahn, R. and Richards, K. (1989) 'The internationalisation of environmental regulation'. *Harvard International Law Journal*, **30**(2) (Spring), pp. 421–46.

Hahn, R. and Stavins, R. (1993) 'Trading in greenhouse permits: A critical examination of design and implementation issues', In W. Clark and H. Lee (eds) *Global Climate Policy*. Cambridge, MA: Harvard University Press.

Hahn, R. and Stavins, R. (1995) 'Trading in greenhouse permits: A critical examination of design and implementation issues'. In H. Lee (ed.) *Shaping National Responses to Climate Change: A Post-Rio Guide*. Washington and California, Island Press, pp. 177–219.

Haigh, N. (1996) 'Climate change policies and politics in the European Community'. In T. O'Riordan and J. Jaeger (eds) *The Politics of Climate Change: A European Perspective*. London and New York: Routledge, pp. 155–86.

Hall, K., Kogan, N. and Plaut, J. (1990) 'A business approach to global climate change'. *International Environmental Affairs*, **2**(4) (Fall), pp. 198–302.

Hall, S. (1977) 'Culture, the media and the ideological effect'. In J. Curran, M. Gurevitch and J. Woollacott (eds) *Mass Communication and Society*. London: Edward Arnold, pp. 315–49.

Hall, S. (1982) 'The rediscovery of "ideology": return of the repressed in media studies'. In M. Gurevitch, et al. (eds) *Culture, Society and the Media*. London: Routledge, pp. 56–91.

Ham, C. and Hill, M. (1984) *The Policy Process in the Modern Capitalist State*. Sussex: Wheatsheaf.

Hamer, M. (1987) *Wheels Within Wheels: A Study of the Roads Lobby*. London: Kegan Paul.

Hamer, M. (1994) 'Head-on collision over transport'. *New Scientist*, 12 November, p. 32.

Hamer, M. and MacKenzie, D. (1995) 'Brussels blocks Britain's clean air plan'. *New Scientist*, 18 November, p. 14.

Hamilton, M. (1998) 'Oil executives are shifting their stance'. *Washington Post*, 3 March, p. 1.

Hampson, F. C. (1990) 'Climate change: Building international coalitions of the like-minded'. *International Journal*, XLV (Winter), pp. 36–76.

Hampson, C. (1992) 'CO$_2$ emissions policy and industry'. In P. Vellinga and M. Grubb (eds) *Climate Policy in the European Community*. Report of October 1992 Workshop. London: RIIA, pp. 48–55.

Hanna, J. (1995) 'Towards a single carbon currency'. *New Scientist*, 29 April, pp. 50–1.

Hannigan, J. A. (1995) *Environmental Sociology: A Social Constructionist Perspective*. London and New York: Routledge.

Hansen, A. (1991) 'The media and the social construction of the environment'. *Media, Culture and Society*, 13, pp. 443–58.

Hansen, A. (1993a) 'Introduction'. In A. Hansen (ed.) *The Mass Media and Environmental Issues*. Leicester: Leicester University Press, pp. xv–1.

Hansen, A. (1993b) 'Greenpeace and press coverage of environmental issues'. In A. Hansen (ed.) *The Mass Media and Environmental Issues*, Leicester: Leicester University Press, pp. 150–79.

Hansen, A. (ed.) (1993c) *The Mass Media and Environmental Issues*. Leicester: Leicester University Press.

Hare, B. (1990) Editorial, *ECO*, issue 8, SWCC, Geneva.

Hare, B. (1992) 'The Climate Convention, What does it mean?'. *ECO*, UNCED Special Issue, June.

Hart, D. M. and Victor, D. G. (1993) 'Scientific elites and the making of US policy for climate change research 1957–74'. *Social Studies of Science*, 23, pp. 643–80.

Hatch, M. (1993) 'Domestic politics and international negotiations: The politics of global warming in the United States'. *Journal of Environment and Development*, 2(2) (Summer), pp. 1–39.

Hatch, M. (1995) 'The politics of global warming in Germany'. *Environmental Politics*, 4(3), pp. 415–41.

Haufler, V. (1995) 'Crossing the boundary between public and private: International regimes and non-state actors'. In V. Rittberger (ed.) *Regime Theory and International Relations*. Oxford: Clarendon Press, pp. 94–112.

Haughland, T. and Roland, K. (1991) 'Energy and environment in China: Challenges ahead'. *International Challenges*, 11(2), pp. 34–40.

Hawkins, A. (1993) 'Contested ground: International environmentalism and global climate change'. In R. Lipschutz and K. Conca (eds) *The State and Social Power in Global Environmental Politics*. New York: Columbia University Press, pp. 221–46.

Hayes, P. and Smith, K. (eds) (1992) *The Global Greenhouse Regime: Who Pays?* London: Earthscan.

Hempel, L. (1993) 'Greenhouse warming: The changing climate in science and politics'. *Political Research Quarterly*, 46(1) (March), pp. 213–40.

Herman, E. S. (1986) 'Gatekeeper versus propaganda models: A critical American perspective'. In M. Golding, G. Murdock and P. Schlesinger (eds) *Communicating Politics*, pp. 171–95.

Herman, S. and Chomsky, N. (1988) *Manufacturing Consent: The Political Economy of the Mass Media*. New York: Pantheon.

Herrick, C. and Jamieson, D. (1995) 'The social construction of acid rain' *Global Environmental Change*, 5(2), pp. 105–12.

Hertsgaard, M. (1996) 'Who's afraid of global warming? Surprise! Its big business that's worried now'. *Washington Post*, 21 January, p. 12.

Highfield, R. (1993) 'Eye in the sky maps the sea by degrees'. *Daily Telegraph*, 22 December, p. 14.

Hildyard, N. (1993) 'Foxes in charge of the chickens'. In W. Sachs (eds) *Global Ecology: A New Arena of Political Conflict*. London and New York: Zed Books, pp. 22–35.

Hlobil, P. (1996) Climate Network Europe, Czech Republic. Interview, Geneva, COP2, 19 July.

Hofrichter, J. (1990) 'Evolution of environmental attitudes in the European Community'. *Scandinavian Political Studies*, 12(2), pp. 119–46.

Hohnen, P. (1991) *ECO*, issue 2, Nairobi.

Hollingsworth, M. (1986) *The Press and Political Dissent*. London: Pluto.

Hope, C., Anderson, J. and Wenman, P. (1993) 'Policy analysis of the greenhouse effect: An application of the PAGE model'. *Energy Policy*, March, pp. 327–37.

Hopgood, S. (1998) *American Foreign Environmental Policy and the Power of the State*. Oxford: Oxford University Press.

Houghton, J. (1994) 'Scientists' View of Global Warming'. *Daily Telegraph*, 23 March, p. 14.

Humphreys, D. (1996) 'Regime theory and NGOs: The case of forest conservation'. In D. Potter (ed.) *NGOs and Environmental Policies: Asia and Africa*. London and Oregon: Frank Cass, pp. 90–109.

Huq, D. (1996) BCAS (Bangladesh Centre for Advanced Studies), Questionnaire on the Politics of Global Warming.

Hurrell, A. (1992) 'Brazil and the international politics of Amazonian deforestation'. In A. Hurrell and B. Kingsburg (eds) *The International Politics of the Environment: Actors and Institutions*. Oxford: Clarendon Press, pp. 398–430.

Hurrell, A. (1995) 'International society and the study of regimes: A reflective approach'. In V. Rittberger (ed.) *Regime Theory and International Relations*. Oxford: Clarendon Press, pp. 49–73.

Hurrell, A. and Kingsbury, B. (eds) (1992) *The International Politics of the Environment: Actors and Institutions*. Oxford: Clarendon Press.

Hyder, T. O. (1992) 'Climate negotiations: The North/South perspective'. In I.M. Mintzer (ed.) *Confronting Climate Change: Risks Implications and Responses*. Cambridge: Cambridge University Press, pp. 323–37.

ICCP (undated) Briefing, 'ICCP comment on future negotiations of climate change issue'.

ICCP (1994, 1995), Letters to Rafe Pommerance, Assistant Secretary at the Environment and Development Department, 19 August 1994, 16 January 1995.

ICCP (1995) Letter to Eileen Claussen, Director of Global Environmental Affairs and Deputy Assistant to the President, 15 March.

ICCP (1996) 'ICCP supports US call for long-term focus on climate change issue'. Press release, 17 July.

Impact (1994) 'Indicators for successful technology transfer', no. 12 (March), p. 14.

International Business (1992) Joint statement presented to INC 5, New York, 18–28 May.

International Chamber of Commerce (ICC) (1992) *Business Brief*, no. 4, (June) 'Energy and protection of the atmosphere'. Prepared for UNCED.

International Chamber of Commerce (ICC) (1995) 'Statement by the International Chamber of Commerce', presented at COP1, Berlin, March 29.

International Chamber of Commerce (ICC) (1996) 'Statement by the International Chamber of Commerce', presented at COP2, Geneva, 8–19 July.

International Climate Change Partnership (1995a) ICCP statement to the US House of Representatives Subcommittee on Energy and Power, House Commerce Committee, 21 March.

International Climate Change Partnership (1995b) 'ICCP commends results of international climate talks: urges business leadership on technical assessment'. Press release, 12 April.

Impact (1994) 'Study of the IPCC greenhouse gas inventory methodology applied to land use and forestry changes in Kenya', no. 12, (March) p. 14.

International Federation of Industrial Energy Consumers (IFIEC) (1996) Letter to Michael
    Zammit Cutajar on climate change and the Second Conference of the Parties, 10 July.
IPCC (1990) *First Assessment Report*, vol. I, overview. Geneva: WMO/UNEP.
IPCC (1995) *Second Assessment Report*, vol. I, overview. Geneva: WMO/UNEP.
Irwin, A. and Davies, H. (1997) 'Landslides strike as US feels heat of global warming'. *Daily
    Telegraph*, 9 October, p. 19.
Isaksen, I. S. A. (1993) 'The Role of scientific assessments on climate change and ozone depletion
    for negotiations of international agreements'. *International Challenges*, 13(2), pp. 76–85.
Ishiumi, Y. (1996) Deputy-Director General, Global Environmental Affairs, Ministry of
    International Trade and Industry, Japan. Interview, Geneva, COP2, 19 July.
Ivins, M. (1998) 'Global warming "debate" is just a red-hot herring'. *Star-Telegram* (USA), 12
    August.
Jachtenfuchs, M. (1996) 'A methodology for frame analysis'. Chap. 3 in *International Policy-
    Making as a Learning Process? The European Union and the Greenhouse Effect*. Aldershot:
    Avebury, pp. 43–60.
Jackson, T. (1995) 'Joint implementation and cost-effectiveness under the Framework Convention
    on Climate Change'. *Energy Policy*, 23(2) (February), pp. 117–38.
Jacobson, H. J. and Weiss, E. D. (1995) 'Strengthening compliance with international environ-
    mental accords: Preliminary observations from a collaborative project'. *Global Governance*, 1,
    pp. 119–48.
Jaeger, J. and Ferguson, H. L. (1991) *Climate Change: Science, Impacts and Policy: Proceedings of
    the Second World Climate Conference*. Cambridge, New York and Melbourne: Cambridge
    University Press.
Jaeger, J. and O'Riordan, T. (1996) 'The history of climate change science and politics'. In
    T. O'Riordan and J. Jaeger (eds) *Politics of Climate Change: A European Perspective*. London
    and New York: Routledge, pp. 1–32.
Jakobsen, S. F. (1997) 'Transnational NGO activity, international opinion, and science – crucial
    dynamics of developing country policy-making on climate'. Paper presented at the conference
    on 'Non-state actors and authority in the global system', Warwick University, 31 October–1
    November.
Jensen, K. B. (1992) 'The politics of polysemy: television news, everyday consciousness and poli-
    tical action'. In P. Scannell, P. Schlesinger and C. Sparks. (eds) *Culture and Power*. London:
    Sage, pp. 218–39.
Jessop, B. (1990) *State Theory: Putting Capitalist States in their Place*. Cambridge: Polity.
Jochem, E. and Hoymeyer, O. (1992) 'The economics of near-term reductions in greenhouse
    gases'. In I.M. Mintzer (ed.) *Confronting Climate Change: Risks, Implications and Responses*.
    Cambridge: Cambridge University Press, pp. 217–37.
Jönsson, C. (1995) 'Cognitive factors in explaining regime dynamics'. In V. Rittberger (ed.) *Regime
    Theory and International Relations*. Oxford: Clarendon Press, pp. 202–23.
Jorgensen, D. W and Wilcoxen, P. J. (1995) 'The economic effects of a carbon tax'. In H. Lee (ed.)
    Shaping National Responses to Climate Change: A Post-Rio Guide. Washington and California:
    Island Press, pp. 237–61.
Kamieniecki, S. (ed.) (1993) *Environmental Politics in the International Arena: Movements, Parties,
    Organisations and Policy*. New York: State University Press.
Karaczum, Z. M. (ed.) (1996) *Independent NGO Evaluations of National Plans for Climate Change
    Mitigation: Central and Eastern Europe*, second review, June. Brussels: CNE.
Karliner, J. (1994) 'The environment industry: Profiting from pollution'. *The Ecologist*, 24(2)
    (March/April), pp. 59–63.
Keeley, J. (1990) 'Toward a Foucauldian analysis of international regimes'. *International
    Organisation*, 44(1) (Winter), pp. 83–105.

Kellogg, W. K. (1987) 'Mankind's impact on climate: The evolution of an awareness'. *Climatic Change*, 10, pp. 113–36.

Kempton, W. (1991) 'Lay perspectives on global climate change'. *Global Environmental Change*, 1 (June), pp. 183–208.

Keohane, R. (1982), 'The demand for international regimes', In S. D. Krasner (ed.), *International Regimes*. Ithaca, NY: Cornall University Press.

Keohane, R. (1984) *After Hegemony: Cooperation and Discord In the World Political Economy*. Princeton NJ: Princeton University Press.

Keohane, R. (1995) 'The analysis of international regimes: Towards a Euro-American research programme'. In V. Rittberger (ed.) *Regime Theory and International Relations*. Oxford: Clarendon Press, pp. 23–49.

Keohane, R. and Milner, H. (eds) (1996) *Internationalisation and Domestic Politics*. Cambridge: Cambridge University Press.

Keohane, R. and Nye, J. (eds) (1972) *Transnational Relations and World Politics*. Cambridge, MA: Harvard University Press.

Keohane, R. and Nye, J. S. (1977) *Power and Interdependence: World Politics in Transition*. Boston: Little Brown.

Khaleel, M. (1996) Deputy Director of Environmental Affairs, Ministry of Planning, Human Resources and the Environment, Maldives. Interview, Geneva, COP2, 17 July.

Khatib, H. (1996) 'Conference report on the sixteenth World Energy Congress of the World Energy Council, Tokyo 1995'. *Energy Policy*, 24(3), pp. 275–7.

Khosla, A. (1980) 'Information is the basis of policy'. *Intermedia* 8(3) (May), pp. 12–15.

Kiernan, V. (1994) 'US trades green points in Bohemia'. *New Scientist*, 7 May, p. 32.

Kinley, R. (1995) 'Communication and review under the Framework Convention on Climate Change'. In M. Grubb and D. Anderson (eds) *The Emerging International Regime for Climate Change*. London: RIIA, pp. 45–51.

Kinrade, P. (1996) Sustainable Cities and Industries Campaign Coordinator, Australian Conservation Foundation, Questionnaire on the Politics of Global Warming.

Kirby, A. (1994) Environment Editor with BBC Radio and Television. Interview, London, 25 February.

Kjellen, B. (1994) 'A personal assessment'. In I. M. Mintzer and J. Leonard (eds) *Negotiating Climate Change*. Cambridge: Cambridge University Press, pp. 149–75.

Knudsen, O. (1979) 'Capabilities, issue-areas and inter-state power: In K. Goldmann and G. Sjöstedt (eds) *Power, Capabilities, Interdependence: Problems in the study of International Influence*. London: Sage, pp. 85–115.

Kranjc, A. (1996) The Hydrometeorlogical Institute's advisor to the Environment Ministry, Slovenia. Interview, COP2, Geneva, 15 July.

Krasner, S. D. (ed.) (1983) *International Regimes*. Ithaca, NY: Cornell University Press.

Krasner, S. (1995) 'Power politics, institutions, and transnational relations'. In T. Risse-Kappen (ed.) *Bringing Transnational Relations Back In: Non-state Actors, Domestic structures and International Institutions*. Cambridge: Cambridge University Press.

Kratochwil, F. (1995) 'Contract and regimes: Do issue specificity and variations of formality matter?'. In V. Rittberger (ed.) *Regime Theory and International Relations*. Oxford: Clarendon Press, pp. 73–94.

Kratochwil, F. and Mansfield, E. (1994) *International Organisation: A Reader*. New York: HarperCollins.

Krause, F. (1997) 'The costs and benefits of cutting US carbon emissions: A critical review of economic arguments and studies used in the media campaigns of US status quo stakeholders'. IPSEP, October.

Kripps, H. (1990) 'Power and resistance'. *Philosophy of the Social Sciences*, 20(2), pp. 170–82.

Kuehls, T. (1996) *Beyond Sovereign Territory: The Space of Ecopolitics*, Borderlines, vol. 4. Minneapolis and London: University of Minnesota Press.

Kuik, O., Peters, P. and Schrijver, N. (1994) *Joint Implementation to Curb Climate Change: Legal and Economic Aspects*. Dordrecht: Kluwer.

Kumar, K. (1991a) 'Mass Media and the Environment: Critical Perspectives'. *Vritta Vidya* (University of Poona), pp. 10–11.

Kumar, K. (1991b) 'Mass Media and Environment Awareness: A Research Perspective'. *Vritta Vidya* (University of Poona), pp. 3–4.

Kydd, A. and Snidal, D. (1995) 'Progress in game-theoretical analysis of international regimes'. In V. Rittberger (ed.) *Regime Theory and International Relations*. Oxford: Clarendon Press, pp. 112–39.

Lacey, C. and Longman, D. (1993) 'The press and public access to the environment and development debate'. *Sociological Review*, 41(2), pp. 207–343.

Lacey, M. (1991) *Government and Environmental Politics*. Washington, DC: Woodrow Wilson Center Press.

Lang, K. and Lang, G. E. (1991) 'Watergate: An exploration of the agenda-setting process'. In M. McCombs and M. Protess (eds) *Agenda-Setting: Readings on Media, Public Opinion and Policy-making*. New Jersey: Lawrence Erlbaum, pp. 277–87.

Lashof, D. (1993) 'Statement for the record on the administration's new policy on global climate change', Senior Scientist NRDC, 26 May.

Lashof, D. (1996) Natural Resources Defence Council (NRDC), United States. Interview, COP2, Geneva, 18 July.

Lazzaro, C. (1993) 'Road from Rio: The role of the media in protecting the environment'. Conference paper, IIC, Mexico City, 21 September.

Lee, H. (ed.) (1995) *Shaping National Responses to Climate Change: A Post-Rio Guide* Washington and California: Island Press.

Leggett, J. (ed.) (1990) *Global Warming: The Greenpeace Report*. Oxford: Oxford University Press.

Leggett, J. (1991) 'Energy and the new politics of the environment'. *Energy Policy*, March, pp. 161–70.

Leggett, J. (1995a) 'Climate industry and the insurance industry: Solidarity among the risk community?' London: Greenpeace UK.

Leggett, J. (1995b) 'Climate change and the financial sector'. *Journal of the Society of Fellows, Chartered Insurance Institute*, 9(2) pp. 119–41.

Leggett, J. (1996) Head of Climate Campaign, Greenpeace UK, Questionnaire on the Politics of Global Warming.

Leggett, J. (1997) 'War of the world'. *Guardian*, 12 November, pp. 4–5.

Leipzig Declaration (1996) 'European and American scientists warn against "premature" actions on global warming'. Press release by the Science and Environmental Policy Project, 10 July.

Levy, D. (1997) 'Business and international environmental treaties: ozone depletion and climate change'. *California Management Review*, 39(3), pp. 54–71.

Levy, D. and Egan, D. (1998) 'Capital contests: National and transnational channels of corporate influence on the climate change negotiations'. *Politics and Society*, 26(3), pp. 337–61.

Levy, M., Keohane, R. and Haas, P. (1993) 'Improving the effectiveness of international environmental institutions'. In P. Haas, R. Keohane and M. Levy (eds) *Institutions for the Earth: Sources of Effective Environmental Protection*. Cambridge, MA: MIT Press, pp. 397–426.

Liberatore, A. (1994) 'Facing global warming: The interactions between science and policy-making in the European Community'. In M. Redclift and T. Benton (eds) *Social Theory and the Global Environment*. London and New York: Routledge, pp. 190–205.

Liefferlink, J., Lowe, P. and Mol, P. (eds) (1993) *European Integration and Environmental Policy*. London: Belhaven Press.

Lindblom, C. (1977) *Politics and Markets: The World's Political-Economic Systems*. New York: Basic Books.

Lindzen, R. S. (1991) 'Summary: Reasons for questioning global warming predictions', pp. 2–6, Presentation at Massachusets Institute of Technology, Cambridge MA, 7 June.

Linne, O. (1993) 'Professional practice and organisation: environmental broadcasters and their sources'. In A. Hansen (ed.) *The Mass Media and Environmental Issues*. Leicester: Leicester University Press, pp. 69–81.

Lipschutz, R. and Conca, K. (1993) *The State and Social Power in Global Environmental Politics*. New York: Columbia University Press.

List, M. and Rittberger, V. (1992) 'Regime theory and international environmental management'. In A. Hurrell and B. Kingsbury (eds) *The International Politics of the Environment: Actors and Institutions*. Oxford: Clarendon Press, pp. 85–109.

Litfin, K. (1993) 'Eco-regimes: Playing tug of war with the nation-state'. In R. Lipschultz and K. Conca (eds) *The State and Social Power in Global Environmental Politics*. New York: Columbia University Press, pp. 94–119.

Litfin, K. (1994) *Ozone Discourses*. New York: Columbia University Press.

Litfin, K. (1995) 'Framing science: Precautionary discourse and the ozone treaties'. *Millennium* 24(2), pp. 251–77.

Lovins, A. and Lovins, L. H. (1992) 'Least-Cost Climate Stabilisation'. In G. I. Pearman (ed.) *Limiting Greenhouse Effects: Controlling Carbon Dioxide Emissions*. London: John Wiley & Sons, pp. 351–443.

Lowe, P. and Morrison, D. (1984) 'Bad news or good news: Environmental politics and the mass media'. *Sociological Review*, 32(1), pp. 75–90.

Lowi, T. (1964) 'American business, public policy, case studies and political theory'. *World Politics*, 16(4), pp. 677–715.

Lukes, S. (1974) *Power: A Radical View*. London: MacMillan.

Lukes, S. (ed.) (1994) *Power*. Oxford: Blackwell.

Lunde, L. (1990) 'Environmental NGO's and the Bergen process: Co-optation or enhanced influence?'. *International Challenges*, 10(2), pp. 24–31.

Lunde, L. (1991a) 'Global warming and a system of tradable emissions permits: A review of the current debate'. *International Challenges*, 11(3), pp. 15–28.

Lunde, L. (1991b) 'Science and politics in the greenhouse. How robust is the IPCC consensus?' *International Challenges*, 11(1), pp. 48–55.

Lunde, L. (1991c) 'Science or Politics in the Global Greenhouse? A study of the development towards scientific consensus on climate change'. Report no. 8. Oslo: Fridjof Nansen Institute, pp. 1–165.

Lunde, L. (1992) 'Towards cost-effective climate agreements: Is a clearinghouse mechanism the logical first step?' *International Challenges*, 12(2), pp. 23–32.

Lunde, L. (1995) 'Greenhouse burden-sharing after Berlin: Economic ideals and political realities'. In M. Grubb and D. Anderson (eds) *The Emerging International Regime for Climate Change*. London: RIIA, pp. 51–9.

MacKinlay, A. (1996) 'Preaching on pollution'. Letters to the *Guardian*, 23 July, p. 14.

MacMillan, J. and Linklater, A. (eds) (1995) *Boundaries in Question: New Directions in International Relations*. London: Pinter.

Maddox, B. (1994) Environmental Correspondent at the *Financial Times*. Telephone interview, 23 February.

Malnes, R. (1996) 'The formation of belief about risk: The case of the greenhouse theory'. EED Report no. 1. Oslo: Fridjof Nansen Institute.

Manners, I. (1993) 'European cooperation on environmental security: The case of limiting the emission of greenhouse gases'. Paper presented at the conference on 'New Directions in International Relations', Keele University, September.

Marchetti, A. (1996) 'Climate change politics in Italy'. In T. O'Riordan and J. Jaeger (eds) *The Politics of Climate Change: A European Perspective*. London and New York: Routledge, pp. 298–330.

Marchaud, de Montigny, P. (1981) 'The impact of information technology on international relations'. *Intermedia*, 9(6), pp. 12–15.

Markham, A. (1994) 'Some like it hot: Biodiversity and the survival of species'. (pamphlet) Geneva: WWFI.

Martinsen, K. D. (1991) 'The Soviet Union and climate changes'. *International Challenges*, 11(1), pp. 25–33.

Marx, K. and Engels, F. (1973) *Manifesto of the Communist Party*. London: Foreign Languages Press.

Masood, E. and Ochert, K. (1995) 'UN climate change report turns up the heat'. *Nature*, 378, p. 25.

Matláry, J. (1991) 'From the internal energy market to a community energy policy?' *International Challenges*, 11(4), pp. 17–23.

Maxwell, J. and Weiner, S. (1993) 'Green consciousness or dollar diplomacy?' *International Environmental Affairs*, 5(1) (Winter), pp. 19–35.

Mayer, P., Rittberger, V. and Zürn, M. (1995) 'Regime theory: Status of the art and perspectives', In V. Rittberger (ed.), *Regime Theory and International Relations*. Oxford: Clarendon Press.

Mazzey, S. and Richardson, J. (eds) (1993) *Lobbying in the EC*. Oxford: Oxford University Press.

McCombs, M. and Protess, M. (1991) *Agenda-Setting: Readings on Media, Public Opinion and Policy-making*. New Jersey: Lawrence Erlbaum.

McCombs, M. and Shaw, D. (1991) 'The agenda-setting function of mass media'. In M. McCombs and M. Protess, *Agenda-Setting: Readings on Media, Public Opinion and Policy-making*. New Jersey: Lawrence Erlbaum, pp. 17–25.

McCormick, J. (1989) *The Global Environmental Movement*. London: Belhaven Press.

McCormick, J. (1991) *British Politics and the Environment*. London: Earthscan.

McCormick, J. (1993) 'International non-governmental organisations: Prospects of a global movement'. In S. Kamieniecki (ed.) *Environmental Politics in the International Arena*. Albany: New York Press, pp. 131–45.

McCully, P. (1991) 'The case against climate aid'. *The Ecologist*, 21(6), pp. 224–51.

McIlroy, A. J. (1998) 'Global warming threat to 40% of Britain's birds'. *Daily Telegraph*, 3 August, p. 7.

McLaughlin, A. and Jordan, G. (1993) 'The rationality of lobbying in Europe: Why are Euro-groups so numerous and so weak? Some evidence from the car industry'. In S. Mazzey and J. Richardson (1993) *Lobbying in the EC*. Oxford: Oxford University Press, pp. 122–61.

McLeod, J., Becker, L. B. and Brynes, J. E. (1991) 'Another look at the agenda-setting function of the press'. In M. McCombs and M. Protess (eds) *Agenda-Setting: Readings on Media, Public Opinion and Policy-making*. New Jersey: Lawrence Erlbaum, M. Gurevitch and J. Woollacott, pp. 47–59.

McQuail, D. (1977) 'The influence and effects of mass media'. In J. Curran, *Mass Communication and Society*. London: Edward Arnold, pp. 70–95.

McQuail, D. (1986) 'Diversity in political communication: Its sources, forms and future'. In P. Golding, G. Murdock and P. Schlesinger (eds) *Communicating Politics*, pp. 133–49.

McQuail, D. (1993) *Media Performance: Mass Communication and the Public Interest*. London: Sage.

Melucci, A. (1989) *Nomads of the Present: Social Movements and Individual Needs in Contemporary Society*, edited by J. Keane and P. Mier. Philadelphia: Temple University Press.

Mestel, R. (1994) 'Is it curtains for the ozone hole?' *New Scientist*, 21 May, p. 3.

Meyer, A. (1994) 'Commons people given green light'. *Guardian*, 29 July, p. 16.

Meyer, A. and Cooper, T. (1995) 'A recalculation of the social costs of climate change'. Occasional paper, *The Ecologist* series, October.

Micheals, P. J. and Knappenberger, P. C. (1995) 'The satanic gases: Political science of the greenhouse effect'. *Economic affairs*, 16(1) (Winter), pp. 19–26.

Milbrath, L. (1993) 'The world is relearning its story about how the world works'. In S. Kamienecki (ed.) *Environmental Politics in the International Arena: Movements, Parties, Organisations and Policy*. New York: State University Press, pp. 21–41.

Miles, E. (1988) 'Expertise and scientific advice in international resource management measures'. *International Challenges*, 8(3), pp. 9–10.

Miliband, R. (1983) *Class Power and State Power*. London: Verso.

Milner, H. (1992) 'International theories of cooperation among nations: Strengths and weaknesses'. *World Politics*, 44(3) (April), pp. 446–96.

Mintzer, I. M. (ed.) (1992a) *Confronting Climate Change: Risks, Implications and Responses*. Cambridge: Cambridge University Press.

Mintzer, I. (1992b) 'Insurance against the heat trap: Estimating the costs of reducing the risks'. In G. I. Pearman. (ed.) *Limiting Greenhouse Effects: Controlling Carbon Dioxide Emissions*. London: John Wiley and Sons, pp. 83–111.

Mintzer, I. M. and Leonard, J. A. (eds) (1994) *Negotiating Climate Change*. Cambridge: Cambridge University Press.

Mitchell, R. (1984) 'Public opinion and environmental politics in the 1970s and 1980s'. In N. Vig and M. Kraft (eds) *Environmental Policy in the 1980s: Reagan's New Agenda*. California: CQ Press, pp. 51–73.

Mitchell, R. (1991) 'From conservation to environmental movement: The development of the modern environmental lobbies'. In M. Lacey (ed.) *Government and Environmental Politics*. Washington: Woodrow Wilson Centre Press, pp. 81–113.

Moltke, K. V. and Rahman, A. (1996) 'External perspectives on climate change: A view from the United States and the Third World'. In T. O'Riordan and J. Jaeger (eds) *The Politics of Climate Change: A European Perspective*. London and New York: Routledge, pp. 330–46.

Moores, S. (1992) 'Texts, readers and contexts of reading'. In P. Scannell, P. Schlesinger and C. Sparks (eds) *Culture and Power*. London: Sage, pp. 137–58.

Morand-Francis, P. D. (1995) 'Lateral thinking and common measures', in M. Grubb and D. Anderson (eds) *The Emerging International Regime for Climate Change*. London: RIIA, pp. 67–79.

Morgenthau, H. (1979) *Power among Nations*, 5th edn. New York: Knopf.

Moss, R. (1995) 'Avoiding dangerous interference in the climate system: The role of values, science and policy'. *Global Environmental Change*, 5(1), pp. 3–6.

Müller, E. (1993) 'Implementation of the German CO_2 reduction goals'. In P. Vellinga and M. Grubb (eds) *Climate Change Policy in the European Community*. London: RIIA, pp. 32–7.

Mulligan, W. (1996) Global Climate Coalition. Interview, Geneva, COP2, 18, July.

Murdock, G. (1982) 'Large corporations and the control of the communications'. In M. Gurevitch, T. Bennett, J. Curran and J. Woollacott (eds) *Culture, Society and the Media*. London: Routledge, pp. 118–51.

Murdock, G. and Golding, B. (1977) 'Capitalism, Communication and Class Relations'. In J. Curran, M. Gurevitch and J. Woollacott (eds) *Mass Communication and Society*. London: Edward Arnold, pp. 12–44.

Murkoweski, F. H. (1996) 'The UN climate change negotiations in Geneva: Playing poker with America's economic future'. Press release, Senate Committee on Energy and Natural Resources, 18 July.

Nadelmann, E. (1990) 'Global prohibition regimes: The evolution of norms in international society'. *International Organisation*, 44(4), pp. 479–526.

National Power (1991) Letter from National Power to Secretary of State for the Environment (Rt. Hon Michael Heseltine), 19 July.

Narwada, S. (1996) Ministry of Local Government and Environment Fiji. Interview, COP2, Geneva, 16 July.

Network (1994) 'Climate Convention inadequate', February/March.

Neuzil, M. and Kovarik, W. (1996) *Mass Media and Environmental Conflict: America's Green Crusades*. London: Sage.

Newell, P. (1995a) 'The Fossil Fuel Lobbies and the Politics of Global Warming'. Inter-disciplinary Research Network on Environment and Society, Keele University, September.

Newell, P. (1995b) 'Climate of Opinion: The Mass Media and Global Warming'. British International Studies Association's Global Environmental Change Workshop, City University, London, October.

Newell, P. (1995c) 'Politics in a warming world'. *Environmental Politics*, 4(4) (Winter), pp. 276–9.

Newell, P. (1996) 'Climate change in the Mediterranean: The NGO dimension'. In *Sustainable Mediterranean*. Brussels: EC Press, pp. 12–13.

Newell, P. (1997) 'Changing landscapes of diplomatic conflict: The politics of climate change post-Rio'. In F. Dodds (ed.) *The Way Forward: Beyond Agenda 21*. London: Earthscan, pp. 37–47.

Newell, P. (1998) 'Who CoPed out at Kyoto?: An assessment of the Third Conference of the Parties to the Framework Convention on Climate Change'. *Environmental Politics*, 7(2) (Summer), pp. 153–60.

Newell, P. and Paterson, M. (1996) 'From Geneva to Kyoto: The Second Conference of the Parties to the United Nations Framework Convention on Climate Change'. *Environmental Politics*, 5(4), (Winter) pp. 729–35.

Newell, P. and Paterson, M. (1998) 'A climate for business: Global warming, the state and capital'. *Review of International Political Economy*, 5(4) (Winter) pp. 679–704.

New Scientist (1995a) 'Costing calamity'. Letters section, 30 September, p. 32.

New Scientist (1995b) 'Price of a life'. Letters section, 23 September, p. 27.

New York Times (1989) 'Greenhouse sceptics', 13 December, p. 4.

New York Times (1996) 'UN climate report was improperly altered, underplaying uncertainties, critics say', 17 June, p. 6.

Nierenberg, W. A., Jastrow, R. and Feitz, F. (1989) *Scientific Perspectives on the Greenhouse Problem*. Washington: Marshall Institute.

Nilsson, S. and Pitt, D. (1994) *Protecting the Atmosphere: The Climate Change Convention and its Context*. London: Earthscan.

Nitze, W. (1991) 'Criteria for negotiating a greenhouse convention that leads to actual emissions reductions'. *International Challenges*, 11(1), pp. 9–16.

Nitze, W. (1994) 'A failure of Presidential leadership'. In I. M. Mintzer and J. A. Leonard (eds) *Negotiating Climate Change*. Cambridge: Cambridge University Press, pp. 187–201.

Nitze, W., Miller, A. S. and Sand, P. H. (1992) 'Shaping institutions to build new partnerships: Lessons from the past and a vision for the future'. In I. M. Mintzer (ed.) *Confronting Climate Change: Risks, Implications, Responses*. Cambridge University Press, pp. 337–51.

Nohrstedt, S. A. (1993) 'Communicative action in the risk-society: public relations strategies, the media and nuclear power'. In A. Hansen (ed.) *The Mass Media and Environmental Issues*. Leicester: Leicester University Press, pp. 81–105.

Norberg-Bohm, V. and Hart, D. (1995) 'Technological cooperation: Lessons from development experience'. In H. Lee (ed.) *Shaping National Responses to Climate Change: A Post-Rio Guide.* Washington and California: Island Press, pp. 261–89.

Norway Daily (1996) 'Secretive oil lobby', 24 June.

Nurmi, S. (1996) Department of the Environment, Finland. Interview, COP2, Geneva, 15 July.

Nuttal, N. (1994) Environmental Correspondent at *The Times.* Interview, 25 February.

Nye, J. and Keohane, R. (1972) 'Transnational relations and world politics'. In R. Keohane and J. Nye (eds) *Transnational Relations and World Politics.* Cambridge, MA: Harvard University Press, pp. ix–xxiv.

Nye, J. and Keohane, R. (1973) 'Transnational relations and world politics: A conclusion'. In N. Goodrich and S. Kay (eds) *International Organisation: Politics and Processes.* Madison: University of Wisconsin Press, pp. 427–54.

Nyirabu, C. (1996) Vice-President's office, Republic of Tanzania. Interview, COP2, Geneva, 17 July.

Observer (1998) 'Man not to blame for global warming', 'Solar win blows away theories', 'So much hot air', 12 April, pp. 1, 9, 24.

Odell, P. (1981) *Oil and World Power,* 6th Edn. London: Penguin.

OECD (1994) *Climate Policy Initiatives,* vol. 1. Paris: OECD.

Ophuls, W. (1977) *Ecology and the Politics of Scarcity: Prologue to a Political Theory of the Steady State.* San Francisco: WH Freeman.

Oppenheimer, M. (1992) 'Lead us to Rio Mr. President'. *New York Times,* 22 February, p. 16.

O'Riordan, T. and Jaeger, J. (eds) (1996) *The Politics of Climate Change: A European Perspective.* London and New York: Routledge.

O'Riordan, T. and Jordan, A. (1996) 'Social institutions and climate change'. In T. O'Riordan and J. Jaeger (eds) *The Politics of Climate Change: A European Perspective.* London and New York: Routledge, pp. 65–106.

O'Riordan, T. and Rowbotham, E. (1996) 'Struggling for credibility: The United Kingdom's response'. In T. O'Riordan and J. Jaeger (eds) *The Politics of Climate Change: A European Perspective.* London and New York: Routledge, pp. 228–68.

Osherenko, G. and Young, O. (1993) 'The formation of international regimes: Hypotheses and cases'. In O. Young and G. Osherenko (eds) *Polar Politics: Creating International Environmental Regimes. Ithaca, NY: Cornell University Press,* pp. 1–21.

Ossewaarde, M. and Mekus, H. (1994) 'In search of further possibilities for joint implementation'. *Change,* 22 November, pp. 5–7.

Ozone Action (1997) 'Ties that blind: Case studies of corporate influence on climate change policy' (paper). Washington, DC: Ozone Action.

Pachauri, R. K. and Bhandari, P. (eds) (1991) 'Global warming: Collaborative study on strategies to limit CO_2 emissions in Asia and Brazil' (paper). New Delhi: Asia Energy Institute.

Pachauri, R. and Bhandari, P. (undated) *Climate Change in Asia and Brazil.* New Delhi: TERI.

Palmer, J. (1992) 'EC wavers over plan to tax carbon dioxide emissions'. *Guardian,* 18 February, p. 12.

Pape, R. (1994) 'Negotiations Stalled'. *Acid News,* 4 October, p. 19.

Parikh, J. and Gokarn, S. (1993) 'Climate change and India's energy policy options: New perspectives on sectoral CO_2 emissions'. *Global Environmental Change: Human and Policy Dimensions,* 3(3) (September), pp. 356–65.

Parry, M., Carter, T. R. and Hulme, M. (1996) 'What is a dangerous climate change?' *Global Environmental Change,* 6(1), pp. 1–6.

Parry, M. and Duncan, R. (1995) *The Economic Implications of Climate Change.* London: Earthscan.

Passacandtando, J. and Carothers, A. (1995) 'Crisis?, What crisis? The Ozone Backlash'. *The Ecologist*, 25(1) (Jan/Feb), pp. 5–7.

Paterson, M. (1992a) 'The Convention on climate change agreed at the Rio Conference'. *Environmental Politics*, 1(4), pp. 267–72.

Paterson, M. (1992b) 'Global warming'. In C. Thomas, *The Environment in International Relations*. London: RIIA, pp. 155–95.

Paterson, M. (1993a) 'The politics of climate change after the Earth Summit'. *Environmental Politics*, 2(4), (Winter), pp. 174–90.

Paterson, M. (1993b) 'Global warming: The great equaliser?' *Journal fur Entwicklungspolitik*, 8(3), pp. 217–28.

Paterson, M. (1994) 'Explaining the Climate Change Convention: Global Warming and International Relations Theory'. PhD thesis, Essex University.

Paterson, M. (1995a) 'Radicalising regimes? Ecology and the critique of IR theory'. In J. MacMillan and A. Linklater (eds) *Boundaries in Question: New Directions in International Relations*. London: Pinter, pp. 212–27.

Paterson, M. (1995b) 'Linking the domestic and the international: historical materialism and the politics of global warming'. European Consortium of Political Research paper, April.

Paterson, M. (1996a) 'IR theory: Neo-realism, neo-institutionalism and the Climate Convention'. In J. Vogler and M. Imber (eds) *The Environment and International Relations*. London: Routledge, pp. 59–77.

Paterson, M. (1996b) *Global Warming and Global Politics*. London and New York: Routledge.

Paterson, M. and Grubb, M. (1992) 'The international politics of climate change'. *International Affairs*, 68(2), pp. 293–310.

Patton, P. (1979a) 'Power and norm: notes'. In M. Meaghlan and P. Patton (eds) *Michel Foucault: Power, Truth and Strategy* Sydney: Feral, pp. 59–66.

Patton, P. (1979b) 'Of power and prisons', In M. Meaghan and P. Patton (eds) *Michel Foucault: Power, Truth and Strategy* Sydney: Feral, pp. 109–147.

Patton, P. (1989) 'Taylor and Foucault on power and freedom'. *Political Studies*, 37, pp. 260–76.

Pearce, F. (1992a) 'Last chance to save the planet?' *New Scientist*, 30 May, p. 27.

Pearce, F. (1992b) 'Industrial growth: Is it part of the solution or the root of the problem?' *New Scientist*, 6 June, pp. 23–5.

Pearce, F. (1993) 'Ancient forests muddy global warming models'. *New Scientist*, 27 November, p. 7.

Pearce, F. (1994) 'Frankenstein syndrome hits climate treaty'. *New Scientist*, 11 June, p. 5.

Pearce, F. (1995a) 'North unmoved by drowning isles'. *New Scientist*, 18 February, p. 5.

Pearce, F. (1995b) 'Global funeral in Berlin?' *New Scientist*, 25 February, p. 1.

Pearce, F. (1995c) 'Climate treaty heads for trouble'. *New Scientist*, 18 March, p. 4.

Pearce, F. (1995d) 'Fiddling while the earth warms'. *New Scientist*, 25 March, pp. 3–4.

Pearce, F. (1995e) 'World lays odds on global catastrophe'. *New Scientist*, 8 April, pp. 5–6.

Pearce, F. (1996) 'Cold shoulder for climate research'. *New Scientist*, 25 February, p. 4.

Pearman, G. I. (ed.) (1992) *Limiting Greenhouse Effects: Controlling Carbon Dioxide Emissions*. London: John Wiley & Sons.

Penny, R. (1996) Climate Network Africa, South Africa. Interview, COP2, Geneva, 17 July.

Peters, G. B. (1994) 'Agenda-setting in the European Community'. *Journal of European Public Policy*, 1(1), pp. 9–26.

Plant, G. (1990) 'Institutional and legal responses to global climate change'. *Millennium*, 19(3), pp. 413–28.

Ploman, E. (1980) 'Information as a symbolic environment'. *Intermedia*, 8(3), (May), pp. 13–16.

Polsby, N. (1980) *Community Power and Political Theory: A further look at problems of evidence and inference*. New Haven: Yale University Press.

Porter, G. and Brown, J. (1991) *Global Environmental Politics*. Boulder, CO: Westview Press.

Porter, G. and Brown, J. (1996) *Global Environmental Politics*, 2nd Edn. Boulder, CO: Westview Press.

Potter, D. (ed.) (1996a) *NGOs and Environmental Policies: Asia and Africa*. London and Oregon: Frank Cass.

Potter, D. (1996b) 'NGOs and environmental politics'. In A. Blowers and P. Glasbergen (eds) *Environmental Policy in an International Context*. London: Arnold, pp. 25–49.

Princen, T. (1994) 'NGOs: creating a niche in environmental diplomacy'. In T. Princen and M. Finger, *Environmental NGOs in World Politics*. London: Routledge, pp. 29–48.

Princen, T. and Finger, M. (1994) *Environmental NGOs in World Politics*. London: Routledge.

Princen, T., Finger, M. and Manno, J. (1994) 'Transnational linkages'. In T. Princen and M. Finger, *Environmental NGOs in World Politics*. London: Routledge, pp. 217–37.

Princen, T., Finger, M. and Manno, J. (1995) 'Nongovernmental organisations in world environmental politics'. *International Environmental Affairs*, **7**(1) (Winter), pp. 43–58.

Protess, M., Cook, J. L., Curtin, T. R., Gordon, M. T., Left, D. R. McCombs, M. E. and Miller, P. (1991) 'The impact of investigative reporting on public opinion and Policy-making: Targeting toxic waste'. In M. McCombs and M. Protess (eds) *Agenda-Setting: Readings on Media Public Opinion and Policy-making*. New Jersey: Lawrence Erlbaum, pp. 171–87.

Putnam, R. (1988) 'Diplomacy and domestic politics: the logic of two-level games'. *International Organisation*, **42**(3) (Summer), pp. 427–60.

Radford, T. (1992) 'Earth's future in the balance, say top scientists'. *Guardian*, 12 February, p. 2.

Raghavan, C. (1994) 'Climate change key decisions put off'. *Third World Resurgence*, no. 5 (September), pp. 4–5.

Rahman, A. (1996) BCAS (Bangladesh Centre for Advanced Studies) and CANSEA (Climate Action Network South East Asia). Interview, COP2, Geneva, 19 July.

Rahman, A. and Roncerel, A. (1994) 'A view from the ground up'. In M. Mintzer and J. A. Leonard (eds) *Negotiating Climate Change*. Cambridge: Cambridge University Press, pp. 239–77.

Ramakrishna, K. (1990) 'North–South issues, common heritage of mankind and global climate change'. *Millennium*, **19**(3), pp. 429–45.

Ramakrishna, K. and Young, O. R. (1992) 'International organisations in a warming world: Building a global climate regime'. In I. M. Mintzer (ed.) *Confronting Climate Change: Risks, Implications, Responses*. Cambridge: Cambridge University Press, pp. 253–65.

Ratnasiri, J. (1996) Environment Division, Ministry of Transport, Environment and Women's Affairs, Sri Lanka. Interview, COP2, Geneva, 16 July.

Raustiala, K. (1996) 'Non-state actors'. In D. Sprinz and U. Luterbacher (eds) *International Relations and Global Climate Change*, PIK report, no. 21. Potsdam: Potsdam Institute, pp. 55–61.

Raustiala, K. (1997) 'States, NGOs, and international environmental institutions'. *International Studies Quarterly*, **41**, pp. 719–40.

Rayner, S. (1993) 'Prospects for CO_2 emissions reductions policy in the USA'. *Global Environmental Change*, **3**(2) (March), pp. 12–31.

Read, P. (1994) *Responding to Global Warming: The Technology and Politics of Sustainable Energy*. London and New York: Zed Books.

Reason, L. (1980) 'Friends of the Earth do a communication audit'. *Intermedia* **8**(3) (May), pp. 34–7.

Reazuddin, M. (1996) Deputy Director, Department of the Environment, Bangladesh. Interview, COP2, Geneva, 17 July.

Redclift, M. and Benton, T. (eds) (1994) *Social Theory and the Global Environment*. New York and London: Routledge.

Reinstein, R. (1996) Canadian Electricity Association, (former head of US delegation to INC meetings under US President Reagan). Interview, COP2, Geneva, 18 July.

Rennie, D. (1998) 'Global warming damages part of Barrier Reef'. *Daily Telegraph*, 22 April.

Reynolds, P. A. and McKinlay, R. D. (1979) 'The concept of interdependence: Its uses and misuses'. In K. Goldmann and G. Sjöstedt (eds) *Power, Capabilities, Interdependence: Problems in the Study of International Influence*. London: Sage, pp. 141–67.

Ribot, J. (1993) 'Market–state relations and environmental policy: Limits of state capacity in Senegal'. In R. Lipschutz and K. Conca (eds) *The State and Social Power in Global Environmental Politics*. New York: Columbia University Press, pp. 24–46.

Richardson, B. (1992) 'Climate change: Problems of law-making'. In A. Hurrell and B. Kingsbury (eds) *International Politics of the Environment: Actors and Institutions*. Oxford: Clarendon Press, pp. 166–83.

Richardson, K. and Corner, J. (1992) 'Reading reception: Mediation and transparency in viewers' reception of TV programme'. In P. Scannell, P. Schlesinger and C. Sparks (eds) *Culture and Power*. London: Sage, pp. 158–82.

Ringius, L. (1997) 'Environmental NGOs and regime change: The case of ocean dumping of radioactive waste'. *European Journal of International Relations*, 3(1), pp. 61–104.

Risse-Kappen, T. (ed.) (1995) *Bringing Transnational Relations Back In: Non-state Actors, Domestic Structures and International Institutions*. Cambridge: Cambridge University Press.

Rittberger, V. (ed.) (1995) *Regime Theory and International Relations*. Oxford: Clarendon Press.

Robertson, T. (1994) Head Political Unit, Greenpeace UK. Telephone interview, 28 April.

Robinson, J. (1996) Eddison Electric Institute (Global Climate Coalition member), United States. Interview, COP2, Geneva, 15 July.

Rosenbaum, W. A. (1974) *The Politics of Environmental Concern*. New York: Praeger.

Rosengren, K. E. (1981) 'Mass media and social change: Some current approaches'. In P. Katz and M. Szecskö (eds) *Mass Media and Social Change*. London: Sage, pp. 189–213.

Rouse, J. (1994) 'Power/knowledge'. In G. Gutting (ed.) *The Cambridge Companion to Foucault*. Cambridge: Cambridge University Press, pp. 92–115.

Rowbotham, E. (1996) 'Legal obligations and uncertainties under the Climate Change Convention'. In A. O'Riordan and J. Jaeger (eds) *The Politics of Climate Change: A European Perspective*. London and New York: Routledge, pp. 32–51.

Rowlands, I. (1992) 'Environmental issues in world politics'. In J. Bayliss and N. Rengger (eds) *Dilemmas of World Politics: International Issues in a Changing World Order*. Oxford: Clarendon Press. pp. 287–308.

Rowlands, I. (1995) *The Politics of Global Atmospheric Change*. Manchester: Manchester University Press.

Royal Society (1998) 'Global environmental change, the public and the media'. Briefing of a workshop, London, 6 May.

Royston, M. (1980) 'The evolution of environmental consciousness'. *Intermedia*, 8(3) (May), pp. 18–21.

Rucht, D. (1993) 'Think globally, act locally? Needs, forms and problems of cross-national cooperation among environmental groups'. In D. Liefferlink, P. Lowe and P. Mol (eds) *European Integration and Environmental Policy*. London: Belhaven Press, pp. 75–95.

Rüdig, W. (1993) 'Dimensions of public concern over global warming: A comparative analysis of West European public opinion'. Paper presented at the 4th Global Warming International Conference.

Rüdig, W. (1995) 'Public opinion and global warming'. Strathclyde papers on government and politics no. 101. Glasgow: University of Strathclyde.

Rusel-Jones, R. (1995) 'Too hot to handle'. *Guardian*, 5 April, p. 24.

Russell, B. (1946) *Power: A New Social Analysis*. London: George Allen and Unwin.

Russell, E. (1994) 'The international environmental crisis: Australia's response'. Working paper 6, Monash University, March. Melbourne: Monash University.

Ryan, S. (1994) Environmental Correspondent at *The Sunday Times*. Interview, London, 2 March.

Sachs, W. (ed.) (1993) *Global Ecology: A New Arena of Political Conflict*. London and New York: Zed Books.

Sale, T. (1996) World Energy Council. Interview, COP2, Geneva, 18 July.

Sand, P. (1991) 'International cooperation: The environmental experience'. In J. Tuchman-Matthews (ed.) *Preserving the Global Environment: The Challenge of Shared Leadership*. New York: W. W. Norton. pp. 236–79.

Sands, P. (ed.) (1993) *Greening International Law*. London: Earthscan.

Santaholma, J. (1995) 'Adequacy debate: The precautionary approach also applies to economy'. *Climate Change Bulletin*, issue 6, 1st quarter, pp. 4–5.

Santaniello, N. (1997) 'Dangers of global warming detailed'. *Florida Sun-Sentinel*, 29 April, p. 1.

Saurin, J. (1996) 'International relations, social ecology and the globalisation of environmental change'. In J. Vogler and M. Imber (eds) *The Environment and International Relations*. London: Routledge, pp. 77–99.

Scannell, P. (1992) 'Public service broadcasting and modern public life'. In P. Scannell, P. Schlesinger and C. Sparks (eds) *Culture and Power*, London: Sage, pp. 317–49.

Scannell, P., Schlesinger, P. and Sparks, C. (eds) (1992) *Culture and Power*. London: Sage.

Schattschneider, E. (1960) *The Semi-Sovereign People: A Realists' View of Democracy in America*. New York: Holt, Rinehart and Wilson.

Schelling, T. (1974) *The Strategy of Conflict*. London: Oxford University Press.

Schillo, B., Giammaerelli, L., Keely, D., Swanson, S. and Wilcoxen, P. (1992) 'The distributional impacts of a carbon tax'. Energy Policy Branch, Environmental Protection Agency, 4 August.

Schlesinger, P. (1992) 'From production to propaganda?' In P. Scannell, P. Schlesinger and C. Sparks (eds) *Culture and Power*. London: Sage, pp. 293–317.

Schmidheiny, S. and the Business Council for Sustainable Development (1992) *Changing Course*. Cambridge, MA: MIT Press.

Schmidt, K. (1991) 'Industrial countries' responses to global climate change'. Environmental and Energy Study Institute, special report, 1 July, Washington, DC.

Schneider, S. (1989) 'The greenhouse effect: Science and policy'. *Science*, 243(2892), pp. 771–85.

Schoenmaeckers, B. (1994) 'Joint Implementation is a step on the road to sustainable development'. *Change*, 22 (November), pp. 16–18.

Schoon, N. (1992) 'Millions could go hungry due to global warming'. *The Independent*, 20 May, p. 3.

Schoon, N. (1993) 'Our green record is not all black'. *The Independent*, 6 December.

Schoon, N. (1994a) Environment Editor at *The Independent*. Interview, London, 25 February.

Schoon, N. (1994b) 'Greenpeace has "firm evidence" of global warming'. *The Independent*, 2 June, p. 6.

Schoon, N. (1994c) 'Millions at risk in expanding dust bowl'. *The Independent*, 13 June, p. 3.

Schoon, N. (1994d) 'Melting Antarctic sounds alarm for globe'. *The Independent*, 24 June, p. 7.

Schoon, N. (1997) 'Developed nations look to a free market in pollution'. *The Independent*, 9 December, p. 5.

Schwartz, P., Collyns, N., Hamik, K. and Henri, J. (1992) 'Modifying the mandate of existing institutions: Corporations'. In I. M. Mintzer (ed.) *Confronting Climate Change: Risks, Implications, Responses*. Cambridge: Cambridge University Press, pp. 281–92.

Sebenius, J. (1991) 'Designing negotiations towards a new regime: The case of global warming'. *International Security*, 15(4) (Spring), pp. 110–48.

Sebenius, J. (1992) 'Challenging conventional explanations of international cooperation: negotiation analysis and the case of epistemic communities'. *International Organisation*, 46(1) (Winter), pp. 230–55.

Sebenius, J. (1994) 'Towards a winning climate coalition'. In I. M. Mintzer and J. A. Leonard (eds) *Negotiating Climate Change*. Cambridge: Cambridge University Press, pp. 227–321.

Senate Committee on Energy and Natural Resources (US) (1995) 'US gets hoodwinked at Berlin Climate Convention'. Press release, 7 April.

Seymour-Ure, C. (1974) *The Political Impact of the Mass Media*. London: Sage.

Shackleton, R. et al. (1992) 'The efficiency of carbon tax revenues'. EPA paper, 2 June.

Shackley, S. (1994) 'Global climate science and policy-making: Multiple studies, reduced realities'. CSEC paper series 94.27. Lancaster University: Centre for the Study of Environmental Change.

Shackley, S. (1997) 'The Intergovernmental Panel on Climate Change: Consensual knowledge and global politics'. *Global Environmental Change*, 7(1), pp. 77–9.

Shackley, S. (undated) 'Global climate change and modes of international science and policy'. Lancaster University: Centre for the Study of Environmental Change.

Shackley, S. (undated) 'Reducing the uncertainty: striking a currency for the emerging constitution of the global environment, science and policy'. Lancaster University: Centre for the Study of Environmental Change.

Shackley, S. and Wynne, B. (1995) 'Integrating knowledges for climate change'. *Global Environmental Change*, 5(2) pp. 113–26.

Shanahan, J. (1993) 'Television and the cultivation of environmental concern: 1988–92'. In A. Hansen (ed.) *The Mass Media and Environmental Issues*. Leicester: Leicester University Press, pp. 181–98.

Sharma, R. (1996) Associate Director, Centre for Science and the Environment, New Delhi, India. Questionnaire on the Politics of Global Warming.

Shell Australia (1996) 'Shell takes up greenhouse challenge'. Media release, 6 June.

Sheppard, F. (1992) 'Global warming could bring disaster to low ports'. *Daily Mail*, 10 August.

Shlaes, J. (1995) 'Developing countries escape climate treaty negotiations with no new obligations'. Press Briefing by Global Climate Coalition, 2 April.

Shlaes, J. (1996) Global Climate Coalition. Interview, COP2, Geneva, 18 July.

Shue, H. (1992) 'The unavoidability of justice'. In A. Hurrell and B. Kingsbury (eds) *The International Politics of the Environment: Actors and Institutions*. Oxford: Clarendon Press, pp. 373–98.

Sieghart, A. (1996) Climate Action Network-UK. Interview, COP2, Geneva, 17 July.

Sierra Club (1997) Letter to Frederico Pena, Secretary of Energy, US Department of Energy, 17 November.

Silberschmidt, G. (1996) International Society of Doctors for the Environment. Interview, COP2, Geneva, 18 July.

Singer, F. (1991) 'No scientific consensus on greenhouse warming'. *Wall Street Journal*, 23 September.

Singer, F. (1992a) 'Earth summit will shackle the planet, not save it'. *Wall Street Journal*, 19 February, p. 12.

Singer, F. (1992b) 'The greenhouse debate continued: An analysis and critique of the IPCC climate assessment'. Science and Environment Policy Project paper, Arlington VA, May, pp. 1–9.

Singer, F. (1992c) 'Warming theories need warming label'. *The Bulletin of the Atomic Scientists*, June, pp. 34–9.

Singer, F. (1992d) 'Rio's results recycled'. *Washington Times*, 3 July, p. 14.

Singh, G. (1994) *Southern NGOs Conference on Climate Change: Conference Proceedings*. Quezon City, Philippines: CANSEA.

Singh, G. (1996) CANSEA (Climate Action Network South East Asia). Interview, COP2, Geneva, 17 July.

Skjaerseth, J. B. (1992) 'The successful ozone negotiations: Are there any lessons to be learned?' *Global Environmental Change*, 2(4), pp. 292–300.

Skjaerseth, J. B. (1994) 'The climate policy of the EC: Too hot to handle?' *Journal of Common Market Studies*, 32(1), pp. 25–45.

Skocpol, T. (1985) 'Bringing the state back in: Strategies of analysis in current research'. In P. B. Evans, D. Rueschemeyer and T. Skocpol, *Bringing the State Back In*. Cambridge, New York and Melbourne: Cambridge University Press, pp. 3–37.

Skolnikoff, E. (1990) 'The policy gridlock on global warming'. *Foreign Policy*, **79** (Summer), pp. 77–93.

Sluijs, J. van der, Eijndhoven, J. van, Shackley, S. and Wynne, B. (1998) 'Anchoring devices in science for policy: The case of consensus around climate sensitivity'. *Social Studies of Science*, 28(2) (April), pp. 291–323.

Smith, M. (1992) 'Melting polar caps will flood coastal towns in 40yrs'. *Daily Telegraph*, 10 August, p. 2.

Smith, S. (1993) 'The environment on the periphery of international relations: An explanation'. *Environmental Politics*, 2(4), pp. 28–45.

SMMT (Society of Motor Manufacturers and Traders) (1990) 'The motor industry and the greenhouse effect', April. London: SMMT.

SMMT (Society of Motor Manufacturers and Traders) (1992) *Royal Commission on Environmental Pollution Transport and the Environment: SMMT evidence*, November. London: SMMT.

SMMT (Society of Motor Manufacturers and Traders) (1993) *Road Transport and Climate Change*. Response by the SMMT to the Department of the Environment's discussion document 'Our national programme for CO_2 emissions', April. London: SMMT.

Snidal, D. (1985a) 'Coordination versus prisoner's dilemma: Implications for international cooperation and regimes'. *The American Political Science Review*, 79, pp. 923–41.

Snidal, D. (1985b) 'The limits of hegemonic stability theory'. *International Organisation*, 39(4), pp. 579–614.

Solesbury, W. (1976) 'The environmental agenda: An illustration of how situations may become political issues and issues may demand responses from government: or how they may not'. *Public Administration*, **54** (Winter), pp. 379–97.

Sparks, C. (1992) 'The popular press and political democracy'. In P. Scannell, P. Schlesinger and C. Sparks (eds) *Culture and Power*. London: Sage, pp. 278–93.

Spector, B. and Korula, A. (1993) 'Problems of ratifying international environmental agreements: overcoming initial obstacles in the post-agreement negotiation process'. *Global Environmental Change*, 3(4) (December), pp. 369–83.

Spencer, M. (1996) Atmosphere Campaigner, Greenpeace UK. Questionnaire on the Politics of Global Warming.

Sprinz, D. and Vaahtoranta, T. (1994) 'The interest-based explanation of international environmental policy'. *International Organisation*, **48**(1) (Winter), pp. 77–105.

Stairs, K. and Taylor, P. (1992) 'Non-Governmental Organisations and the Legal Protection of the Oceans: A Case Study'. In A. Hurrell and B. Kingsbury (eds) *The International Politics of the Environment: Actors and Institutions*. Oxford: Clarendon Press, pp. 110–41.

Stanford, A. (1996) Energy Campaigner, Friends of the Earth UK. Questionnaire on the Politics of Global Warming.

Stavins, R. (1993) 'Transaction costs and the performance of markets for pollution control' Paper presented at Harvard University, 8 December.

Steen, N. (ed.) (1994) *Sustainable Development and the Energy Industries: Implementation and Impacts of Environmental Legislation*. London: RIIA/ Earthscan.

Stein, A. (1983) 'Coordination and collaboration: Regimes in an anarchic world'. In S. D. Krasner (ed.) *International Regimes*. Ithaca, NY: Cornell University Press, pp. 115–40.

Stein, A. (1990) *Why Nations Cooperate: Coordination and Choice in International Relations.* Ithaca, NY: Cornell University Press.

Stirpe, D. and Fay, K. (1995) 'Final report from the first meeting of the Conference of the Parties to the Framework Convention on Climate Change'. Memo to the International Climate Change Partnership, 17 April.

Stoll, R. and Ward, M. D. (1989) *Power in World Politics*. Boulder: Lynne Rienner.

Stram, B. N. (1995) 'A carbon tax strategy for global climate change'. In H. Lee (ed.) *Shaping National Responses to Climate Change: A Post-Rio Guide.* Washington and California: Island Press, pp. 1–41.

Strange, S. (1983) 'Cave! hic dragones: A critique of regime analysis'. In S. Krasner (ed.) *International Regimes*. Ithaca, NY: Cornell University Press, pp. 337–54.

Strange, S. (1988) *States and Markets: An Introduction to International Political Economy.* London: Pinter.

Strange, S. (1994) 'Rethinking structural change in the international political economy: States, firms and diplomacy'. In R. Stubbs and G. Underhill (eds) *Political Economy and the Changing Global Order.* Basingstoke: MacMillan, pp. 103–16.

Strange, S. (1996) *The Retreat of the State.* Cambridge: Cambridge University Press.

Studelska, J. and Buchanan, B. (1993) *Summary Report of The Freedom Forum.* Environmental Journalism Summit, 9–10 January.

Sunday Express (1991) 'Experts got it wrong over global warming', 20 December, p. 4.

Sunday Telegraph (1992) 'Plague of caterpillars linked to global warming', 5 July, p. 3.

Sunday Telegraph (1994) 'More hot air than facts on global warming', 26 June, p. 2.

Sung, J. (1997) Sppech to China Council for International Cooperation on Environment and Development, 5 October.

Supertran, A. D. (1996) Department of the Environment and Natural Resources, Philippines. Interview, COP2, Geneva, 16 July.

Susskind, L. (1994) *Environmental Diplomacy: Negotiating more effective global agreements.* Oxford: Oxford University Press.

Susskind, L. and Ozawa, O. (1992) 'Negotiating more effective international environmental agreements'. In A. Hurrell and B. Kingsbury (eds) *The International Politics of the Environment: Actors and Institutions.* Oxford: Clarenden Press, pp. 142–65.

Swart, R. J., de Boois, H. and Rotmans, J. (1989) 'Targeting climate change' *International Environmental Affairs*, 1(3) (Summer), pp. 222–34.

Sydnes, A. K. (1991) 'Global climate negotiations: Another twenty years of fruitless North–South bargaining?' *International Challenges*, 11(1), pp. 58–65.

Sydnes, A. K. (1996) 'Norwegian climate policy: Environmental idealism and economic realism'. In T. O'Riordan and J. Jaeger (eds.) *The Politics of Climate Change: A European Perspective.* London and New York: Routledge, pp. 268–98.

Tanabe, A. and Grubb, M. (1991) 'The greenhouse effect in Japan: Burden or opportunity'. In M. Grubb, P. Brackley, M. Ledic, A. Mathur, S. Rayner, J. Russell and A. Tanabe, *Energy Policies and the Greenhouse Effect. Volume Two: Country Studies and Technical Options.* London: RIIA/Dartmouth, pp. 279–317.

Taplin, R. (1994) 'Greenhouse Policy Development and the Influence of the Climate Change Prediction Timetable: An Australian Perspective'. Paper presented at the conference on Time-Scales and Environmental Change, South–North Centre for Environmental Policy, SOAS, London, 23 April.

The Independent (1992) 'Global warming could lengthen the hayfever season', 21 May, p. 4.

The Independent (1994) 'Sunspots linked to global warming', 7 April, p. 7.

The Independent (1995a) 'Global warming could boost apple harvest', 4 January, p. 14.

The Independent (1995b) 'Australia abandons clean air target', 11 February, p. 14.

The Times (1992) 'Eye in the sky on ice', 20 April, p. 4.

The Times (1993) 'Pollution charges put luxury cars at risk', 7 March, p. 2.

The Times Magazine (1993) 'Theories on ice', 11 December, pp. 35–6.

Thomas, C. (1992) *The Environment in International Relations*. London: RIIA.

Tolba, M. (1989) 'A step-by-step approach to protection of the atmosphere'. *International Environmental Affairs*, 1(4), pp. 304–9.

Tuchman-Matthews, J. (ed.) (1991) *Preserving the Global Environment: The Challenge of Shared Leadership*. New York: W.W. Norton.

Turner, L. (1978) *Oil Companies in the International System*. London: George Allen and Unwin.

Uezono, M. (1996) Japanese Citizens Alliance for Saving the Atmosphere and Earth (CASA), Japan. Interview, COP2, Geneva, 19 July.

Underdal, A. (1989) 'The politics of science in international resource management: A summary'. In S. Andresen and W. Ostreng (eds) *International Resource Management: The Role of Science and Politics*. New York and London: Belhaven Press, pp. 253–69.

Underdal, A. (1995) 'The study of international regimes'. *Journal of Peace Research*, 32(1), pp. 113–19.

Ungar, S. (1992) 'The rise and (relative) decline of global warming as a social problem'. *The Sociological Quarterly*, 33(4), pp. 483–501.

UNICE (1994) 'Report of the task force on the CO_2 tax and voluntary agreements', 16 February.

United Nations Framework Convention on Climate Change (UNFCCC) (1992). New York: United Nations.

United Nations General Assembly (UNGA) A/AC.237/37 (9 July 1993); A/AC.237/37/Add.1 (12 July 1993); A/AC.237/37/Add.3 (14 July 1993); A/AC.237/37/Add.4 (14 July 1993); A/AC.237/35. (20 July 1993); A/AC.237/37/Add.2 (29 July 1993).

US Senate (1996) Letter to US President Bill Clinton, 10 July.

Vatikiotis, M. (1994) 'For profit's sake: Western utilities promote conservation in Sabah'. *Far Eastern Economic Review*, 14 April, p. 14.

Vaughan, D. and Mickle, C. (eds) (1993) *Environmental Profiles of European Business*. London: RIIA/Earthscan.

Vaus de, D. A. (1991) *Surveys in Social Research*, 3rd edn. London: Allen and Unwin/UCL Press.

Vellinga, P. and Grubb, M. (eds) (1993) *Climate Change Policy in the European Community*. Report of October 1992 Workshop. London: RIIA.

Vidal, J. (1992) 'Going the way of Atlantis'. *Guardian*, 1 May, p. 15.

Vidal, J. (1994) 'Why some lives are cheaper than others'. *Guardian*, 1 July, pp. 12–13.

Vig, N. J. and Kraft, M. E. (1984) *Environmental Policy in the 1980s. Reagan's New Agenda*. Washington: Congressional Quarterly Inc.

Villagrasa, D. (1998) 'The microscope COP/Slow foxtrot instead of a tango'. COP4 summary, posted on 'Can-talk', 15 December.

Vogler, J. (1995) *The Global Commons: A Regime Analysis*. Sussex: John Wiley and Sons.

Vogler, J. (1996) 'Introduction', in Volger and Imber (1996), pp. 1–22.

Vogler, J. and Imber, M. (eds) (1996) *The Environment and International Relations*. London: Routledge.

Walcott, C. (1996) European Roundtable of Industrialists. Interview, COP2, Geneva, 18 July.

Wall Street Journal (1996) 'A major deception on global warming', 6 June, p. 16.

Waltz, K. (1979) *A Theory of International Politics*. Reading, MA: Addison-Wesley.

Wapner, P. (1995) 'Politics beyond the state: Environmental activism and world civic politics'. *World Politics*, 47 (April), pp. 311–40.

Wapner, P. (1996) *Environmental Activism and World Civic Politics*. Albany: State University of New York Press.

Ward, H. (1996) 'Game theory and the politics of global warming: The state of play and beyond'. *Political Studies*, XLIV, pp. 850–71.

Weber, M. (1947) *The Theory of Social and Economic Organisation*. London: Routledge and Kegan Paul.

Weir, F. (1992) 'Appeals by electricity generating companies on grounds of commercial confidentiality'. Letter from Friends of the Earth, UK to Rt Hon. Michael Heseltine, Secretary of State for the Environment, 4 February.

Weir, F. (1996) Former Climate Campaigner, Friends of the Earth UK. Questionnaire on the Politics of Global Warming.

Weizsäcker, E. U. Von (1994) *Earth Politics*. London and New Jersey: Zed Books.

Wendt, A. (1987) 'The agent-structure problem in international relations theory'. *International Organisation*, 41(3) (Summer), pp. 335–70.

Wendt, A. (1992) 'Anarchy is what states make of it: social construction of power politics'. *International Organisation*, 46(2), pp. 391–425.

Westergaard, J. (1977) 'Power, class and the media'. In J. Curran, M. Gurevitch and J. Woollacott (eds) *Mass Communication and Society*. London: Edward Arnold, pp. 95–116.

Wettestad, J. (1991) 'Verification of International Greenhouse Agreements: A mismatch between technical and political feasibility?' *International Challenges*, 11(1) pp. 41–7.

Wettestad, J. (1995) 'Nuts and bolts for environmental negotiators? Designing effective international regimes: A conceptual framework', January. Bergen: Fridtjof Nansen Institute.

Whitelegg, J. (1994) 'Road-builders make their pitch'. *New Scientist*, 30 April, pp. 48–9.

Whitney, D. C. and Becker, L. B. (1991) 'Keeping the Gates for Gatekeepers: The effects of wire news' In M. McCombs and M. Protess (eds) *Agenda-Setting: Readings on Media, Public Opinion and Policy-making*. New Jersey: Lawrence Erlbaum, pp. 229–36.

Wiarda, H. J. (1990) *Foreign Policy Without Illusion: How Foreign-Policy-Making Works and Fails to Work in the United States*. Little Brown: Scott, Foresham and Co.

Willetts, P. (ed.) (1982) *Pressure Groups in the Global System: The Transnational Relations of Issue-Orientated Non-governmental organisations*. London: Pinter.

Willetts, P. (1993) *Transnational Actors and Changing World Order*. Occasional Papers Series no. 17. Yokohama: PRIME.

Willetts, P. (1996a) 'Who cares about the environment?'. In J. Vogler and M. Imber (eds) *The Environment and International Relations*. London: Routledge, pp. 120–38.

Willetts, P. (1996b) 'From Stockholm to Rio and beyond: The impact of the environmental movement on the UN consultative arrangements'. *Review of International Studies*, 22(1) (January), pp. 57–81.

Willetts, P. (ed.) (1996c) *The Conscience of the World*. London: C. Hurst.

Wilson, G. (1990) *Business and Politics: A Comparative Introduction*, 2nd edn. London: Macmillan.

Wirth, D. (1989) 'Climate Chaos'. *Foreign Policy*, no. 74 (Spring), pp. 3–22.

Wood, A. (1993) 'The Global Environment Facility Pilot Phase'. *International Environmental Affairs*, 5(3) (Summer), pp. 219–33.

Woollacott, J. (1982) 'Messages and meanings'. In M. Gurevitch T. Bennett, T. Curran and J. Woollacott (eds) *Culture, Society and the Media*. London: Routledge, pp. 91–113.

World Coal Institute (1998) Press release on kyoto Protocol.

World Industry Council for the Environment (undated) Executive brief on climate change.

World Resources Institute (WRI) (1991) *Greenhouse Warming: Negotiating a Global Regime*. Washington: WRI.

World Wide Fund for Nature (1996) 'Intensifying efforts on the Berlin Mandate'. Position statement, June.

Worthington, S. (1992) 'Beaches to vanish as sea rises 3ft'. *The Sun*, 11 August, p. 1.

Wrong, D. H. (1979) *Power: Its Uses, Forms and Bases*. Southampton: Basil Blackwell.

Wynne, B. (1994) 'Scientific knowledge and the global environment'. In M. Redclift and T. Benton (eds) *Social Theory and the Global Environment*. New York and London: Routledge, pp. 169–90.

Wynne, B. (1995) 'Modelling global climate: The epistemic and the political'. Paper presented at Keele University, Sociology Department Seminar, 6 March.

Yamin, F. (1995) 'Additional commitments and joint implementation: the post-Berlin landscape'. In M. Grubb and D. Anderson (eds) *The Emerging International Regime for Climate Change*. London: RIIA, pp. 59–67.

Young, O. (1986) 'International regimes: Toward a new theory of institutions'. *World Politics*, 39(1) (October), pp. 104–22.

Young, O. (1989a) 'Science and social institutions: Lessons for international resource regimes'. In S. Andresen and W. Ostreng (eds) *International Resource Management: The Role of Science and Politics*. New York and London: Belhaven Press, pp. 7–25.

Young, O. (1989b) 'The politics of international regime formation: managing natural resources and the environment'. *International Organisation*, 43(3) (Summer), pp. 349–75.

Young, O. (1989c) *International Cooperation: Building Regimes for Natural Resources and the Environment*. Ithaca, NY: Cornell University Press.

Young, O. (1998) *Global Governance: Learning lessons from the environmental experience*. Cambridge, MA: MIT Press.

Young, O., Demko, G. and Ramakrishna, K. (1991) *Global Environmental Change and International Governance*. Summary and recommendations of a conference held at Dartmouth College. Paper, Dartmouth College, Manover, NH.

Young, O. and Osherenko, G. (eds) (1993) *Polar Politics: Creating International Environmental Regimes*. Ithaca, NY: Cornell University Press.

Young, O. and Von Moltke, K. (1994) 'The consequences of international environmental regimes'. *International Environmental Affairs*. 6(4) (Fall), pp. 348–68.

Young, S. C. (1993) *The Politics of the Environment*. Manchester: Baseline Books.

Zukin, S. and Dimmagio, P. (eds) (1990) *Structures of Capital: The Social Organisation of the Economy*. Cambridge: Cambridge University Press.

Zürn, M. (1995) 'Bringing the second image (back) in: About the domestic sources of regime formation'. In V. Rittberger (ed.) *Regime Theory and International Relations*. Oxford: Clarendon Press, pp. 282–315.

Index